Blue Light Responses:

Phenomena and Occurrence in Plants and Microorganisms

Volume II

Editor

Horst Senger, Dr. rer.nat.
Professor of Botany
Department of Biology/Botany
Philipps-University
Marburg
West Germany

CRC Press, Inc.
Boca Raton, Florida

Library of Congress Cataloging-in-Publication Data

Blue light responses.
  Includes bibliographies and indexes.
  1. Blue light—Physiological effect. 2. Plants,
Effect of blue light on. 3. Ultraviolet radiation —
Physiological effect. I. Senger, H. (Horst), 1931-
QH515.B569 1987   574.19′154   86-8325
ISBN 0-8493-5235-5 (v. 1)
ISBN 0-8493-5236-3 (v. 2)

This book represents information obtained from authentic and highly regarded sources. Reprinted material is quoted with permission, and sources are indicated. A wide variety of references are listed. Every reasonable effort has been made to give reliable data and information, but the author and the publisher cannot assume responsibility for the validity of all materials or for the consequences of their use.

Direct all inquiries to CRC Press, Inc., 2000 Corporate Blvd., N.W., Boca Raton, Florida, 33431.

© 1987 by CRC Press, Inc.

International Standard Book Number 0-8493-5235-5 (Volume I)
International Standard Book Number 0-8493-5236-3 (Volume II)

Library of Congress Card Number 86-8325
Printed in the United States

THE EDITOR

Horst Senger, Dr. rer.nat., is Professor of Botany, Department of Botany, Philipps-University Marburg.

He received his doctoral degree from the University of Göttingen and his Habilitation from the University of Marburg. He has held postdoctoral positions at the University of Tübingen and at Oregon State University and has been Guest Scientist at the University of Natal, University of Tokyo, Oregon State University, and at the CSIRO in Canberra.

Professor Senger has worked on synchronization of microalgae, chloroplast development, chlorophyll biosynthesis, photohydrogen evolution, and blue light effects on algae. His work has been published in over 120 original papers, and he is co-editor of a book *Regulation of Chloroplast Differentiation*. He has organized two International Conferences on the "Effect of Blue Light on Plants and Microorganisms" and edited two books, *The Blue Light Syndrome* and *Blue Light Effects in Biological Systems*.

CONTRIBUTORS

Volume I

Helga Drumm-Herrel, Dr. rer.nat.
Akademischer Oberrat
Biological Institute II
University of Freiburg
Freiburg
West Germany

Roger Durand, Dr.
Fungal Differentiation Laboratory
University of Lyon I
Villeurbanne
France

Donat P. Häder, Dr.
Dozent
Department of Biology
Philipps University
Marburg
West Germany

Wolfgang Kowallik, Dr. rer.nat.
Professor
Faculty of Biology
University of Bielefeld
Bielefeld
West Germany

Hans Mohr, Dr. rer.nat.
Professor
Biological Institute II
University of Freiburg
Freiburg
West Germany

Helga Ninnemann, Dr. phil.nat.
Professor
Faculty of Chemistry
University of Tübingen
Tübingen
West Germany

W. Rau, Dr.
Professor
Botanical Institute
University of Munich
Munich
West Germany

Rainer Schmid, Dr. rer.nat.
Dozent
Institute of Plant Physiology
  Cell Biology, and Microbiology
Free University
Berlin
West Germany

Erich L. Schrott, Dr. rer.nat.
Dozent
Botanical Institute
University of Munich
Munich
West Germany

Horst Senger, Dr. rer.nat.
Professor
Department of Biology/Botany
Phillips-University
Marburg, West Germany

CONTRIBUTORS

Volume II

Matthew J. Dring, Ph.D.
Department of Botany
The Queen's University of Belfast
Belfast
Northern Ireland

Paul A. Galland, Dr.
Research Assistant Professor
Department of Physics
Syracuse University
Syracuse, New York

Jonathan B. Gressel, Ph.D.
Professor
Department of Plant Genetics
The Weizman Institute of Science
Rehovot
Israel

Benjamin A. Horwitz, Ph.D.
Postdoctoral Fellow
Department of Plant Biology
Carnegie Institution of Washington
Stanford, California

Christer Larsson, Ph.D.
Assistant Professor
Department of Plant Physiology
University of Lund
Lund
Sweden

Günter Ruyters, Ph.D.
Assistant Professor
Faculty of Biology IV
University of Bielefeld
Bielefeld
West Germany

Werner Schmidt, Dr.
Docent
Faculty of Biology
University of Konstanz
Konstanz
West Germany

Horst Senger, Dr.
Professor
Department of Botany
Philipps-University of Marburg
Marburg
West Germany

Pill-Soon Song, Ph.D.
Professor
Department of Chemistry
University of Nebraska
Lincoln, Nebraska

Susanne Widell, Ph.D.
Department of Plant Physiology
University of Lund
Lund
Sweden

Eduardo Zeiger, Ph.D.
Professor
Department of Biological Sciences
Stanford University
Stanford, California

TABLE OF CONTENTS

Volume I

INTRODUCTION
Chapter 1
Introduction 3
Horst Senger

PHENOMENA, DISTRIBUTION, AND MECHANISMS OF BLUE LIGHT EFFECTS
Chapter 2
Blue Light Effects on Carbohydrate and Protein Metabolism 7
Wolfgang Kowallik

Chapter 3
Photoregulation of Eukaryotic Nitrate Reductase 17
Helga Ninnemann

Chapter 4
A Photosensitive System for Blue/UV Light Effects in the Fungus *Coprinus* 31
Roger Durand

Chapter 5
Blue Light Control of Pigment Biosynthesis—Carotenoid Biosynthesis 43
Werner Rau and Erich L. Schrott

Chapter 6
Blue Light Control of Pigment Biosynthesis—Anthocyanin Biosynthesis............ 65
Helga Drumm-Herrel

Chapter 7
Blue Light Control of Pigment Biosynthesis—Chlorophyll Biosynthesis 75
Horst Senger

Chapter 8
Blue Light Effects in Endogenous Rhythms 87
Rainer Schmid

Chapter 9
Photomovement 101
Donat-P. Häder

INTERACTION WITH OTHER LIGHT EFFECTS
Chapter 10
Mode of Coaction between Blue/UV Light and Light Absorbed by Phytochrome
in Higher Plants 133
Hans Mohr

Chapter 11
The Relation of Photosynthesis to Blue Light Effects 145
Donat-P. Häder

INDEX..163

Volume II

THE MECHANISM OF BLUE LIGHT ACTION

Chapter 1
Possible Primary Photoreceptors.. 3
Pill-Soon Song

Chapter 2
Primary Reactions and Optical Spectroscopy of Blue Light Photoreceptors 19
Werner Schmidt

Chapter 3
Action Spectroscopy...37
Paul A. Galland

Chapter 4
First Measurable Effects Following Photoinduction of Morphogenesis 53
Benjamin A. Horwitz and Jonathan B. Gressel

Chapter 5
Control of Enzyme Capacity and Enzyme Activity.......................................71
Günter Ruyters

Chapter 6
Membrane-Bound Blue Light Receptors — Possible Connection to Blue Light
Photomorphogenesis..89
Suzanne Widell

Chapter 7
Plasma Membrane Purification 99
Suzanne Widell and Christer Larsson

ECOLOGICAL RELEVANCE OF BLUE LIGHT EFFECTS

Chapter 8
Cellular and Functional Properties of the Stomatal Response to Blue Light 110
Eduardo Zeiger

Chapter 9
Marine Plants and Blue Light 121
Matthew J. Dring

Chapter 10
Sun and Shade Effects of Blue Light on Plants 141
Horst Senger

INDEX..153

The Mechanism of Blue Light Action

Chapter 1

POSSIBLE PRIMARY PHOTORECEPTORS

Pill-Soon Song

TABLE OF CONTENTS

I.   Introduction 4

II.   Flavins 4
        A.   Spectroscopy 4
        B.   Photochemical Processes: Possible Implications 6
                1.   Intramolecular Reactions 6
                2.   Intermolecular Reactions 6

III.   Carotenoids 9
        A.   Spectroscopy 9
        B.   Photochemical Processes: Possible Implications 10

IV.   Chromophores Other Than Flavins and Carotenoids 11

V.   Conclusions: Speculative Remarks 12

Acknowledgements 15

References 16

I. INTRODUCTION

A sensory or developmental response of an organism to blue-wavelength light (400 to 500 nm) is triggered by the absorption of quanta by photoreceptors. The light-energized photoreceptor then initiates primary molecular reaction(s) that lead to the subsequent transduction chain; in analogy to the enzyme reaction catalyzing a substrate reaction, photoreceptor and quanta can be thought of as catalyst and substrate, respectively.

One of the essential prerequisites for the efficient operation of a photobiological process in light-responsive organisms is that the photoreceptor in its excited state undergoes fast or ultrafast primary reaction(s) that lead to the sensory transduction chain (Figure 1). Such primary processes, fast enough to compete effectively with other relaxation modes of the excited state, would ensure the high photosensitivity of photobiologically responsive organisms. Figure 2 illustrates the role of primary photoreceptors in light-signal perception with flavin as the hypothetical chromophore of blue light receptors.

Unlike the photoreceptors of vision and photomorphogenesis, blue light receptors remain unidentified. The question as to the possible identity of the chromophore(s) of blue light receptors has centered around the choice between flavins and carotenoids. More than 10 years ago the question as to the possible identity of the chromophore(s) of blue light receptors was examined in terms of the photophysical and photochemical reactivities of flavins and carotenoids.[1,2] The main arguments were that flavins are intrinsically more likely to be the blue light photoreceptors than are carotenoids. Subsequently, it was argued that flavins could act as the primary photoreceptor for a number of responses of organisms to blue light and that the short fluorescence lifetime of flavins bound to the corn coleoptile plasmalemma could serve as a photophysical assay for the photoreceptor flavins (cf. Figure 2).[3,4] On the other hand, the photobiological function of carotenoids primarily involves the absorption of blue light as the light-harvesting antenna. The antenna function of carotenoids is facilitated by the binding of carotenoids to proteins and/or membranes and by chromophore-chromophore dipolar (exciton) interactions between and/or among the bound carotenoid molecules.

In the present review, spectroscopic and photochemical properties of flavins and carotenoids will be examined to ascertain whether these chromophores can act as primary blue light receptors. More recently, relative merits of flavin vs. carotenoid as blue light receptors have been reviewed by Presti,[5] suggesting that carotenoids perhaps function as photoreceptors for several cases of blue light-regulated responses and flavins acting as photoreceptors more prominently in blue light phenomena. Schmidt[6] has reviewed various flavin systems as a workable model for blue light effects. The reader is referred to these reviews for extensive literature coverage of the topic discussed in this chapter.

In the present chapter, flavins and carotenoids are reexamined as blue light receptors in terms of their spectroscopic and photochemical properties. Rhodopsin-like photoreceptors are treated here as a separate category from long-chain carotenoids, since their role as visual and energy-transducing photoreceptors for higher animals and *Halobacterium halobium*, respectively, is well established.

II. FLAVINS

A. Spectroscopy

Flavins show three major absorption bands at about 450, 370, and 260 to 270 nm. These bands are attributable to $\pi \rightarrow \pi^*$ type transitions with oscillator strengths of 0.19, 0.20, and 0.66, respectively, although other types of transitions such as $n \rightarrow \pi^*$ may contribute to some of these band intensities either directly by overlapping band or via vibronic mixing between $n \rightarrow \pi^*$ and $\pi \rightarrow \pi^*$ transitions. A number of blue light action spectra in prokaryotic and eukaryotic organisms can be qualitatively matched with the absorption spectra of flavins.

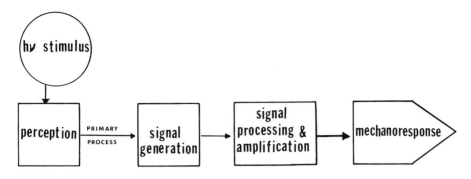

FIGURE 1.   A general scheme for the photosensory transduction of a light-responsive organism. Signal perception by photoreceptor is equivalent to the spectral absorption of stimulus quanta by the chromophore, producing the excited state from which primary process begins.

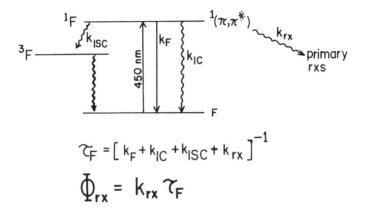

$$\tau_F = \left[k_F + k_{IC} + k_{ISC} + k_{rx} \right]^{-1}$$

$$\phi_{rx} = k_{rx}\,\tau_F$$

FIGURE 2.   The energy level diagram and the primary photoprocess of a flavin (F) photoreceptor (cf Figure 1) for blue light responses in organisms. In general, the primary photoprocess (k_{rx}) competes effectively with fluorescence and radiationless decays (internal conversion and intersystem crossing) so that the excited state lifetime is very short and the efficiency of the primary photoprocess is high, resulting in a highly sensitive response of an organism to light; ϕ_F (quantum yield of fluorescence) $= k_F \cdot \tau_F \simeq 0$ for an active photoreceptor and ϕ_{rx} (quantum yeild of primary reaction) $= k_{rx} \cdot \tau_F \simeq 1$ for an efficient photoreceptor.

The action spectra of blue light responses have been analyzed in detail in terms of quantum-chemically predicted $\pi \to \pi^*$ transitions, yielding an apparent agreement between the observed and calculated spectra of lumiflavin in the near-UV and blue regions.[7,8] However, this analysis cannot be valid, since the theory erroneously assigns the blue absorption band at 450 nm to the second $\pi \to \pi^*$ ($S_0 \to S_2$) transition, their predicted $S_0 \to S_1$ transition being in the near-IR region. Other molecular orbital calculations have reasonably reproduced the observed $\pi \to \pi^*$ transition energies of flavins.[9,10]

To characterize the dichroic orientation of photoreceptors and to explain the possible polarotropic effects of blue light responses, information on the orientation of transition dipoles is essential. The polarized single-crystal absorption of flavin mononucleotide (FMN) in flavodoxin indicates that both S_1 and S_2 bands are polarized along the long molecular axis,[11] making an angle of 20° between the two transition moments. Johansson et al.[12] argue that this angle is in fact 29°. The Kronig-Kramers transformation of specular reflection spectra of crystalline *bis*(10-methylisoalloxazine) copper(II) perchlorate tetrahydrate yields corresponding absorption spectra.[13] Again, S_1 and S_2 transitions are found to be polarized along the long molecular axis, with an angle of 25° between them. Considering this, agreement between the two sets of polarization data is satisfactory. Likewise, earlier theoretical transition moment directions[9,10] calculated from Pariser-Parr-Pople (PPP) (S_1 approximately along C_8-N_3 and S_2 along C_7-C_2) are generally in good agreement with the above experiment. Polarization directions of S_1 and S_2 bands are not significantly altered by substituting N_5 with carbon, as in deazaflavin, yielding an angle of 25° (from fluorescence polarization) and 27° between the two transitions (from molecular orbital calculation[14]). For the tabulation of polarizaiton angles between S_1 and S_n ($n = 2,3$) transitions deduced from different techniques and workers, the reader is referred to Schmidt.[15]

B. Photochemical Processes: Possible Implications

As has been discussed previously,[1] flavins exhibit relatively long-lived fluorescence and phosphorescence emissions, indicating that their excited singlet (S_1) and triplet (T_1) states are sufficiently stable to be in their energized, metastable electronic structure for long enough to initiate various molecular processes during their lifetimes. For efficient blue light response of an organism, the excited state from which the primary molecular process will most likely initiate is the singlet, rather than triplet excited state, since the intersystem crossing is relatively inefficient in flavins (cf. Figure 2), particularly in flavins bound to proteins/membrane.[3,4] The nature of the primary molecular processes (k_{rx} in Figure 2) is not known, but an order of magnitude estimate of k_{rx} has been calculated to be 10^{10} sec^{-1} for the possible flavin photoreceptor in the paraflagellar body of *Euglena gracilis*.[16] In the following paragraphs, several photochemical processes are enlisted for consideration as possible primary molecular processes of flavin photoreceptors.

1. Intramolecular Reactions

It is well known that flavins undergo photodegradation reactions which involve oxidation of the ribityl side chain by the excited isoalloxazine nucleus of flavins.[17—22] The photoreactivity of the flavin nucleus arises from the increased electron affinity and the pKa changes of the excited state relative to the ground state.[23] Examples of the photodegradation reactions of flavins(F) are shown below:

$$^1F(\text{singlet excited state}) \qquad\qquad \rightarrow \text{lumichrome} \qquad (1)$$

$$^3F(\text{excited triplet state}) \qquad\qquad \rightarrow 2'\text{-oxoflavin} \qquad (2)$$

$$^3F \xrightarrow{\qquad\qquad\qquad\qquad} \qquad\qquad \text{FMF} \qquad (3)$$

where FMF stands for 7,8-dimethyl-9-(formylmethyl)isoalloxazine. Since these reactions are self-destructive, it seems unlikely that they play an important role in the primary molecular processes of blue light responses in organisms. However, this reasoning does not necessarily rule out the role of an intramolecular photoreaction in the primary process. For example, Reaction 2 can be coupled with an enzymatic reduction of the 2'-oxoribityl side chain to reversibly regenerate the original flavin photoreceptor.

2. Intermolecular Reactions

Flavins are efficient in sensitizing a number of reactions of different substrates. Flavin-photosensitized reactions may proceed via the triplet and/or singlet-excited flavin, depending on the photosensitization mechanisms involved. In general, the oxidative reactions (Type I and II) are sensitized by the flavin triplet state; in Type II oxidation,

$$^3F + {}^3O_2 \rightarrow F + {}^1O_2 \tag{4a}$$

$$1_{O_2}(\text{singlet oxygen}) + (\text{substrate}) \rightarrow S_{\text{oxidized}} \tag{4b}$$

Alternatively (Type I),

$$^3F + S \rightarrow FH^- + S^{\ddagger} \tag{5a}$$

$$S^{\ddagger} + {}^3O_2 \rightarrow S_{\text{oxidized}} \tag{5b}$$

Whether or not these reactions are relevant to the primary photoprocess of blue light receptors is ambiguous. However, it is interesting to observe that photosensitizing dyes such as erythrosin, methylene blue, and eosin provoke photoresponses (e.g., photophobic response) in *Parmaecium* in the presence of light stimulus absorbed by the dyes.[24,25] Type II mechanism of photosensitization with phycocyanins as photosensitizer has been proposed to play a role in the sensory transduction chain of cyanobacteria *Anabaena*.[26] Although evidence in the literature for the direct role of photosensitized oxidations is not available, flavins are exceptionally suited as blue light-absorbing sensitizers, as they generate singlet molecular oxygen with high efficiency.[27] One of the major arguments not favoring the photosensitization as a viable primary photoprocess is that oxidizing agents, particularly singlet oxygen, are freely diffusible through cellular fluids, proteins, and membranes and they are often non-specific relative to the location and the kind of substrates to be oxidized, contrary to the general view that the primary photoreceptors are localized in membranes and their primary reactions are coupled to a specific transduction chain.

Perhaps the most likely type of reaction for the primary molecular processes of blue light receptors is photooxidation or electron transfer reactions without molecular oxygen as the primary oxidizing agent. Light-induced electron transfer is well known in the primary reaction of chlorophylls in photosynthesis. Possible relevance of this type of reaction (illustrated below) to the blue light reception has been discussed elsewhere:[1]

$$^1F + IAA(\text{indole-3-acetic acid}) \rightarrow F\text{-}IA(4a\text{-adduct}) + CO_2 \tag{6}$$

$$^1F + PAA(\text{phenylacetic acid}) \rightarrow F\text{-}PA(4a\text{-adduct}) + CO_2 \tag{7}$$

These oxidative decarboxylations of auxins are most likely mediated by the singlet excited state of flavin.[3] Galston[28] originally observed that IAA can be rapidly photooxidized in the presence of riboflavin, and we have shown that the singlet excited state of flavin can be quenched efficiently by IAA, suggesting that Reaction 6 occurs at a diffusion-controlled rate.[3] Although lateral diffusion of IAA is currently favored as the phototropic transduction reaction in higher plants, high quantum yield of the above reactions suggests that the primary photoprocess of flavin photoreceptors may involve similar reaction mechanisms. It is conceivable that auxin oxidation product(s) released from Reaction 6 by aeration act(s) as a competitive ligand to the auxin receptors, replacing auxins from the receptor sites. The reader is referred to Galston's recollection of his earlier work on the flavin-photosensitized decomposition of IAA as a possible mechanism of phototropism.[29]

The most prominent photoreactions sensitized by flavins involve the following type

$$^{1 \text{ or } 3}F + D(\text{electron/hydrogen donor}) + 2 H^+ \rightarrow FH_2 + D_{ox} \tag{8}$$

where D can be, for example, NADH or EDTA as an artificial electron donor.[30] In fact, it is likely, with flavins as the blue light receptors, that the primary photoprocess is an electron

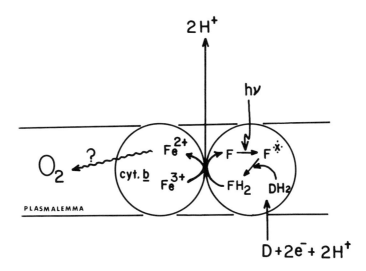

FIGURE 3.   A vectorial flux of protons coupled with the photoinduced elec-
tron/proton transfer of flavin (F) with *b*-type cytochrome in plant plasmalemma, based on the light-induced absorbance changes (LIAC).

transfer from an as-yet unidentified donor to the singlet excited state of flavins. Briggs and
co-workers have accumulated evidence for the possible involvement of the photoinduced
electron transfer reactions between the excited flavin and *b*-type cytochrome in corn coleoptile
phototropism (see review by Senger and Briggs[31] and for subsequent work on light-induced
absorbance changes (LIAC) from Briggs' laboratory (e.g., see *Annual Report of the Director
of Carnegie Institution/Plant Biology Series 1981—1984*). Earlier, Schmidt and Butler[31]
demonstrated in a model system that a cytochrome can be photoreduced by an electron donor
in the presence of flavins. Figure 3 illustrates this type of electron transfer reaction occurring
in coleoptile plasmalemma. A highly speculative mechanism for the flavin-sensitized electron
transfer as a primary photoreaction in blue light-induced movement of bacteria has been
proposed on the basis of a vectorial proton flux,[32] as illustrated in Figure 3. Unfortunately,
the identity and the primary photochemical mechanism of blue light receptors in prokaryotic
and eukaryotic organisms still remain to be established. One can thus argue that the proposal
of the photoinduced electron transfer as the primary photoprocess in blue light responses is
an open question.

   Another photochemical reaction that is potentially important in blue light phenomena,
especially in those cases of blue light-induced transcriptional responses such as carotenogenesis which are relatively slow compared to movement responses such as phototaxis, is
the production of the plant growth regulator ethylene via flavin sensitization

$$^{1 \text{ or } 3}F + \text{methionine} \rightarrow FH_2 + \text{methional(4a-adduct?)} + CO_2 + NH_3 \qquad (9a)$$

$$^{1 \text{ or } 3}F + 2 \text{ methional} \rightarrow FH_2 + 2CH_2{=}CH_2 + 2 \text{ HCOOH} + (CH_3S)_2 \qquad (9b)$$

In addition to methionine, ethionine, cystathione, homocystine, and homocysteine also
undergo photooxidation sensitized by flavins, resulting in varying yields of ethylene.[33] Is
ethylene produced in vivo by blue light irradiation? If so, can it mediate *some* of the blue
light phenomena in plants? These are potentially interesting questions to ponder.

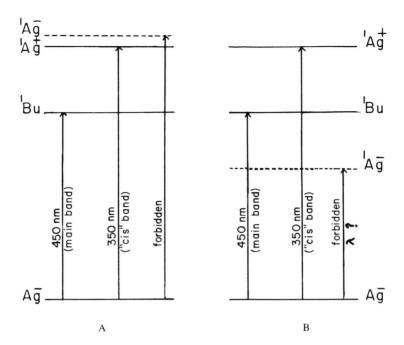

FIGURE 4. The electronic states and $\pi \rightarrow \pi^*$ transitions of β-carotene and related carotenoids based on the traditional (A) vs. new (B) electronic state ordering. In the new state ordering, the lowest state (S_1) is a dipole-forbidden A_g^- state which is electronically covalent in character, whereas the strong, dipole-allowed state (B_u) is at a higher energy level than the A_g^- state. No definitive experimental evidence for the lowest state of A_g^- nature has been observed for long-chain carotenoids.

III. CAROTENOIDS

A. Spectroscopy

The new electronic state ordering is different from the traditional picture. Thus, in the new assignment, the lowest electronically excited state (S_1) is an $^1A_g^-$ state (forbidden and too weak to be clearly observable in symmetric polyenes and carotenoids), in contrast to the allowed ^1B_u state (at 450 nm for β-carotene) as the lowest excited state. Two different orderings of the excited states for β-carotene as a typical carotenoid are illustrated in Figure 4. The location of the forbidden $^1A_g^-$ state (S_1 in the new ordering) and its absorption maximum and intensity are not known at this time. It is predicted that the $A_g \rightarrow ^1A_g^-$ ($S_0 \rightarrow S_1$) absorption occurs in the near IR region (>600 nm).[34] This long-wavelength transition is symmetry forbidden and is not readily observable under the condition of monophoton excitation. However, the loss of molecular symmetry enhances the transition probability and it is then possible to predict such an absorption band. In one case, a weak absorption appears at λ longer than the main B_u absorption band in carotenals.[35] Whether this is attributable to an A_g^- band remains to be seen. The $^1A_g^-$ band is also strongly allowed in two-photon absorption.

In the traditional picture for β-carotene, it is well known that the first transition, $A_g \rightarrow ^1B_u$ (or $A \rightarrow ^1B$ in Platt notation), is strongly electric dipole-allowed with its polarization direction along the long molecular axis. Both oscillator strength ($f \sim 3.4$ for all-*trans* β-carotene) and λ_{max} increase with conjugation. The latter converges to an asymptotic value of 610 nm for "infinitely" long polyenes as a result of C–C and C=C bond-order alterations which restrict the electron motion in a one-dimensional periodic potential and/or π-electron correlation effects.[36]

Both $A_g \rightarrow {}^1A_g^+$ and $A_g \rightarrow {}^1A_g^-$ transitions (Figure 4) are symmetry-forbidden, but the former can be readily observed even in *trans*-polyenes, and it becomes strongly allowed in *cis*-polyenes ("*cis* peak") due to the loss of the center of symmetry.

The assignment of ${}^1A_g^-$ to the lowest excited ${}^1\pi,\pi^*$ state has a profound implication in reinterpreting the spectroscopic and photochemical properties of carotenoids. Recent results further confirm that the ${}^1A_g^-$ state is the lowest excited singlet state for polyenes of up to six conjugated double bonds.[37—41] Recent theoretical calculations predict that the ${}^1A_g^-$ state (vertical) should remain about 4000 cm^{-1} below the allowed ${}^1B_u$ state for large polyenes of $\geqslant 8$ conjugated double bonds.[42] As mentioned earlier, however, the location of the ${}^1A_g^-$ state in longer polyenes, including carotenoids, has not been resolved unambiguously and, consequently, its assignment remains to be established.

A new band has been observed for β-carotene film at 536 and 537 nm for all-*trans* and 15,15'-*cis* isomers, respectively. On the basis of the resonance Raman excitation profile spectrum, Thrash et al.[43] proposed that the $S_0 \rightarrow S_1$ transition in β-carotene is due to a transition to the ${}^1A_g^-$ state as the lowest state. However, the ${}^1A_g^-$ assignment for β-carotene cannot be viewed as definitively established until the possibility of impurities present and produced during spectral measurements has been rigorously dealt with. It is more likely that the ${}^1A_g^-$ band occurs at λ much longer than 536 to 537 nm, possibly at ~800 nm. From luminescence (585 nm) and resonance Raman spectroscopy of crystalline β-carotene powders, a long-wavelength absorption band at about 520 nm has been observed.[44] This absorption band is probably not due to a transition to the ${}^1A_g^-$ state.

B. Photochemical Processes: Possible Implications

The two different electronic state assignments for the lowest S_1 state of carotenoids imply that primary photoprocesses in blue light responses proceed from the ${}^1B_u$ (with ionic character) or ${}^1A_g^-$ (covalent character) state in the traditional or new electronic state ordering, respectively, provided that carotenoid is indeed the primary photoreceptor (see later).

With ${}^1B_u$ as the S_1 state, the primary photoreaction ($\underline{k}_{rx}$) must compete with highly efficient internal conversion (Figure 2). The radiationless relaxation (internal conversion) is so efficient that the mean lifetime (τ) of the ${}^1B_u$ state is predicted to be ~10^{-15} sec from the integration of the absorption band ($\epsilon(\tilde{\nu})$ with maximum $\tilde{\nu}_{max}$ in wavenumber; n is the refractive index)

$$\tau({}^1B_u) = \tau_F^0 \times \Phi_F = [2.9 \times 19^{-9} \, \tilde{\nu}_{max}^2 \, n^2 \int_0^\infty \epsilon(\tilde{\nu}) \, d\nu]^{-1} \times \Phi_F$$

$$\leqslant 1 \times 10^{-9} \times 10^{-6}$$

$$\leqslant 10^{-15} \text{ sec}$$

where τ_F^0 and Φ_F are the radiative fluorescence lifetime and an order-of-magnitude estimate of the fluorescence quantum yield, respectively. Clearly, the ultrashort lifetime is not conducive to action of the carotenoid as a primary photoreceptor with rate constant $\underline{k}_{rx}$, as has been pointed out previously.[2]

On the other hand, with ${}^1A_g^-$ as the lowest excited state (S_1), the blue absorption band is attributable to the allowed $A_g \rightarrow {}^1B_u$ ($S_0 \rightarrow S_2$) transition. The primary photoreaction may then proceed from either the ${}^1B_u$ (S_2) or ${}^1A_g^-$ (S_1) state, if the radiationless transition between these two states is not efficient. This does not seem to be a likely possibility, since the vibronic coupling between the ${}^1B_u$ and ${}^1A_g^-$ states can be strong,[45] suggesting that the relaxation from ${}^1B_u$ to the hypothetical lower state (${}^1A_g^-$) is so fast that a direct primary photoprocess from ${}^1B_u$ for blue light responses is unlikely. If the ${}^1B_u \rightarrow A_g^-$ relaxation is slow, then the assignment of ${}^1A_g^-$ to the S_1 state in carotenoids requires that the blue light photoreceptor with the carotenoid as the chromophore also acts as a red light receptor, especially if the $A_g \rightarrow {}^1A_g^-$ transition is at least partially allowed as a result of the carotenoid molecule being nonsymmetric (e.g., carotenals, carotenones, and their Schiff's bases). Furthermore, the radiative lifetime of the ${}^1A_g^-$ state is expected to be much longer than that of the ${}^1B_u$ state. These factors (nonsymmetric carotenoids and long lifetimes) would favor the action of carotenoids as both blue and red receptors. However, there is no experimental evidence for the dual spectral nature of blue light receptors.

It has been pointed out that the primary photoreceptors for sensory transduction in both prokaryotes and eukaroytes possess covalently linked chromophores.[3] For retinylic photo-receptors, such as rhodopsin and bacteriohodopsin, the covalent linking between the chromophore and protein is provided by the protonated Schiff's base formation.

With carotenals and carotenones, the protonated Schiff's base linkage shifts the blue absorption maximum of these carotenoids toward red,[46] thus eliminating protonated carotenoid Schiff's bases as primary *blue light* receptors. Neutral Schiff's bases show very little spectral shift with respect to the absorption maxima of the parent carotenoids and can spectrally qualify as blue light receptors. However, if the ${}^1A_g^-$ assignment for the lowest S_1 state is assumed, the neutral carotenoid Schiff's bases should also act as red light receptors, in addition to being blue light receptors. Since the spectral duality is not experimentally supported, it seems unlikely that large carotenoids with more than six conjugated double bonds can be considered as the favored photoreceptor molecules as opposed to flavins.

In contrast to flavins, the photochemistry of carotenoids is not diverse. The role of photoisomerization in the primary photoprocesses of rhodopsin and bacteriorhodopsin is well established, and will not be discussed here. However, it is noteworthy that the twisting of double bond(s) in the *cis-trans* isomerization cannot occur from the allowed ${}^1B_u$ excited state with ionic character, for long polyenes with more than six conjugated double bonds.[47] β-Carotene does not photoisomerize in the absence of oxygen or other catalysts such as iodine. Because of the red shift of the allowed ${}^1B_u$ state in rhodopsin and bacteriorhodopsin, relative to the neutral retinal or its Schiff's bases, it is unlikely that the lowest singlet excited state in these photoreceptors remains ${}^1A_g^-$. Thus, in rhodopsin and bacteriorhodopsin, photoisomerization most likely proceeds from the B_u state.

Aside from photoisomerization, both photooxidation and photoreduction of fucoxanthin films in 2 M KOH have been observed under electrochemical conditions on the Pt electrode.[48] However, the reaction conditions used are not suitable to ascertain the nature of the photoreactivity of this carotenoid.

IV. CHROMOPHORES OTHER THAN FLAVINS AND CAROTENOIDS

It has been frequently mentioned that chromophores which are different from flavins and carotenoids may be involved in blue light responses. For example, multiple photoreceptors are apparently involved in mediating phototropism in *Phycomyces blakesleeanus*.[49] In this section, we will consider several blue light absorbers as such possible candidates.

Bilirubin serves as a blue light receptor in the phototherapy of infant jaundice, in which the concentration of toxic bilirubin isomer in circulation is decreased.[50,51] Bilirubin is unusual in that only one major absorption band in the blue region constitutes its UV-visible absorption spectrum. The blue absorption maximum and its intensity (ϵ) are sensitive to solvent polarity.[52] The natural isomer (4-Z, 15-Z) shows a λ_{max} at 438 nm ($\epsilon = 47,200$) in phosphate buffer, pH 7.4,[52] whereas the absorption maximum shifts to 450 nm in chloroform with an increase in the molar extinction coefficient. Upon binding to human serum albumin, bilirubin absorbs maximally at 460 nm ($48,500\ M^{-1}cm^{-1}$; pH 7.4).[53] The photoisomerization of the Z,Z isomer to ''photobilirubin'' isomers (4, Z, 15 E; 4 E, 15 Z) is accompanied by some changes in λ_{max} and ϵ_{max}, but the most significant change occurs in the CD spectra of these isomers.[52]

Some of the other naturally occurring compounds with the absorption spectra similar to the action spectra for blue light responses are listed below for future examination: chrysazine, coleone A, isocoleone A, dihydroflavylium salts, and naphthazarins. It is not known what chromophores are responsible for the 475-nm-absorbing photoreceptor(s) (without near-UV peaks) in plants.[54,55]

V. CONCLUSIONS: SPECULATIVE REMARKS

Flavins exhibit diverse photoreactivities. For example, the $^{1,3}(\pi,\pi^*)$ state flavins can mediate (1) proton and/or electron transfer, and (2) photosensitization of a number of reactions including redox reactions and photorepair of DNA damage (thymine-thymine dimer; see later). They are also involved in chemiluminescent and bioluminescent processes. Some of the well-resolved action spectra of blue light responses[56-58] can thus be understood in terms of the intrinsic photoreactivity of flavins.

Can carotenoids be ruled out as primary photoreceptors? In asking this question, we obviously exclude retinals from the carotenoid category. In this review and many others, carotenoids have usually been referred to β-carotene and its analogs with long enough conjugation to absorb in the blue spectral region. It appears that carotenoids such as β-carotene without a functional group that can covalently link to the apoprotein moiety are not likely to be viable candidates as blue light receptors, since the chromophore relaxation resulting from photoisomerization and/or polarizability changes cannot lead to specific coupling with the transducing membrane/effector molecules. Most recently, it has been shown that a carotenoidless mutant of *Neurospora crassa* (carotenoid content <0.5% of the wild type) has the same threshold for photoinduction of protoperithecia, suggesting that carotenoids are not the blue light receptor.[59] Furthermore, if the lowest excited state is of $^1A_g^-$ character, carotenoids are also expected to act as red light receptors. There is no evidence among the reported action spectra for blue light responses that suggests the involvement of carotenoids acting in the red/far-red region. One might argue that the near-red action peak at 595 nm for phototropic and growth responses of the sporangiophore of *Phycomyces* is attributable to an $^1A_g^-$ excitation of a carotenoid, rather than a flavin triplet excitation.[60] However, evidence is against the former.[60]

Are there carotenoids other than retinals with functional groups that can link to proteins (e.g., via Schiff's base) and that can possibly function as primary blue light receptors? Rhodopsin- and bacteriorhodopsin-like carotenoprotein Schiff's bases (protonated) cannot be reconciled with many blue light action spectra with about 450-nm peaks. However, rhodopsin- and carotenoprotein-neutral Schiff's bases cannot be ruled out *a priori* for blue light responses in aneural organisms, especially for those responses with action spectral maxima shifted to about 500 nm.

In a most interesting report, Foster et al.[61] described the induction of phototactic response in a carotenoidless mutant of *Chlamydomonas* by exogenously added retinal isomers, thus suggesting that retinal or rhodopsin-like chromophore is the photoreceptor for phototaxis in this alga. Since retinals are the photoreceptor chromophore of animal vision and their photochemistry is well established, the above observation lends strong support for the retinylic identity of blue light receptors in this organism. Unfortunately, the method used for phototaxis measurements in the above work does not rule out photodynamic and thermal effects of the exogenous retinals in the alga. As pointed out earlier, *Paramecium* without photoreceptor pigments can respond to light when artificial dyes are incorporated into the cell.[24,25] Furthermore, *Chlamydomonas* swims faster with increasing temperature.[62] It is, therefore, conceivable that the artificially retinal-pigmented *Chlamydomonas* absorbs light of actinic wavelengths and the local photodynamic action of the cell contributes to the "tactic" response described by Foster et al.[61] In *Halobacteria,* there seems to be no doubt

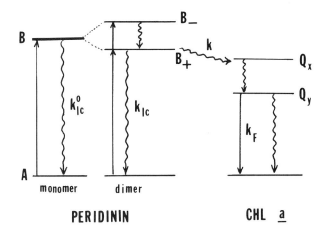

FIGURE 5.   Illustration of energy transfer from carotenoid (peridinin) to chlorophyll *a* in photosynthetic light-harvesting pigment complexes of dinoflagellates. Energy transfer is facilitated by the resonance stabilization of the dipole-allowed B state (B_u in symmetry notation; cf. Figure 4), as the result of exciton interactions between two peridinin molecules (dimer). For more details and references, see elsewhere.[2,3]

that retinal is the chromophore of photosensory receptors for near UV (370 nm; absorbed principally by slow-cycling rhodopsin [sR]) and red (587 nm; absorbed by sR_{587}).[64] However, the *Chlamydomonas* action spectra are contradictory to the *Halobacteria* action spectra, if retinal is indeed the chromophore for the former. It is of course conceivable that the spectral and photochemical properties of the retinylic photoreceptor in *Chlamydomonas* are different from those of *Halobacteria*.

In concluding this chapter, several questions are listed below in lieu of definitive conclusions which cannot be drawn from the present state of knowledge on primary blue light receptors:

1. *Like most other sensory photoreceptors, are the chromophores of blue light receptors covalently linked to apoproteins?* Both flavins and carotenoids with covalent linkages are known in nature. It is interesting that the only established flavin photoreceptor in the native DNA photolyase is apparently covalently linked via a 4a-adduct.[70,72] This adduct with λ_{max} ~ 380 and only weak absorption in the blue is well protected from oxygen, since the flavin chromophore is released only after denaturation. Thus, it is worthwhile to search for the elusive photoreceptors which may be covalently linked.

2. *What is the known role of carotenoids (other than rhodopsin and bacteriorhodopsin) in blue light phenonomena?* Answers to this question have been discussed previously.[3] The author feels that carotenoids such as carotenes are unlikely to be primary photoreceptors for blue light responses that are sensory in nature, as opposed to blue light-induction-transcription responses, on the basis of the spectroscopic and photochemical properties discussed earlier. However, carotenoids play important roles in blue light responses of aneural organisms. For example, blue light-absorbing carotenoids can function as antenna pigments, as in photosynthesis, transferring the absorbed quanta to primary photoreceptors, as illustrated with dinoflagellate light-harvesting pigment proteins in Figure 5.[3] They can function as a screening pigment, for example, to produce spatial light intensity gradients between the lighted and shaded sides of plant tissues.[65] Carotenoids have also been proposed to act as a highly refractive quasicrystalline layer (quarter-wave stack) to focus an actinic beam on to the primary photoreceptors in *Chlamydomonas*.[66]

3. *Are there separate photoreceptors for fast sensory responses and slow induction/ transcription effects of blue light?* If we accept flavins as the likely chromophore for the

former (cf. the second question above), can carotenoids be possibly photoreceptors for the latter? The action spectrum of carotenogenesis in *N. crassa* that has been carefully corrected for the optical bias of mycelial pad lacks a strong near-UV peak characteristic of a flavin absorption spectrum. On this basis and other lines of evidence, carotenoids or other chromophores have been proposed as the primary photoreceptor.[67] It has been pointed out, however, that the extinction of the near-UV absorption depends strongly on flavin derivatives occurring in nature and to some extent on binding environment.[68] Nonetheless, both photoisomerization and photooxidation of carotenoids are possible in the presence of suitable electron acceptors such as oxygen and iodine. While these reactions may not be sufficiently fast to compete with other radiationless processes of long-chain carotenoids, they may produce products which in turn act as inducers or stimulants for induction/transcription processes elicited by blue light. This remains to be tested.

4. *What is the known role of flavins as blue-light receptors?* In spite of the arguments that flavins best qualify as blue light receptors, the nature of primary photoprocesses remains obscure. Although LIACs may well be physiologically relevant to the blue-light phenomena, direct evidence for the involvement of LIAC and flavin photoreceptors is still lacking. At the moment, only one photobiological reaction has been identified with a flavin photoreceptor. Specifically, photoreactivating enzymes (PRE) of *E. coli*,[70,71] *Streptomyces griseus*,[69] and *Saccharomyces cerevisiae*[72] contain a flavin (deazaflavin in bacterial PREs) as photoreceptor for the dissociation of thymine-thymine dimers in *uv*DNA. The mechanism of this reaction is not known, however. More evidence for a direct role of flavins as photoreceptors remains to be established for blue light responses of aneural organisms. Recently, Horwitz et al.[73] have been able to record an "in vivo" cryptochrome spectrum in *Trichoderma*. The cryptochrome as a blue light receptor for the induction of sporulation in this organism exhibits absorbance maxima at 455 and 480 nm. Combination of this type of sensitive absorbance measurements in vivo and the use of mutants may shed more light on the nature of blue light receptors and their primary photoprocesses.

One can also ask whether ethylene can be photoproduced in vivo to account for some of the slow blue light responses. In fact, it is likely that ethylene is produced in vivo nonenzymatically in the presence of blue light. It is thus interesting to investigate if such photoproduction of ethylene is relevant to blue light-induced carotenogenesis and other induction/transcription responses of organisms.

5. *Are rhodopsin-like photoreceptors the mediator of blue-light responses?* The answer appears to be affirmative for near-UV and red light responses in *Halobacteria,* but evidence is still lacking for blue light responses in prokaryotic and eukaryotic cells.

6. *Are there chromophores other than flavins and carotenoids that function as the primary receptors for blue light phenomena?* This is a completely open question. In order to assess the validity of the question itself, it is necessary to ask whether flavins and carotenoids can be ruled out *a priori* as photoreceptors for certain blue light responses in question. Unfortunately, this is not an easy question to answer. However, one possibility is to examine the action spectra over the entire spectral region (UV, near-UV, and blue), as has been done recently for the photoinduced formation of perithecia.[74,75] In this study, it was possible to show that UV radiation at 280 nm was about 1.7 times more effective than blue light at 450 nm, on a relative quantum basis. Thus, such an action spectrum can be taken as evidence for a flavin-like photoreceptor chromophore and against carotenoids, which exhibit only weak absorption at 280 nm, and against possibly other chromophores, especially if corrections for UV screening and UV-induced inhibitory effects can be made.[74,75] Unfortunately, even with satisfactory corrections for the UV bias, one cannot easily ascertain the photoreceptor identity if the apoproteins contain aromatic amino acids which transfer the absorbed UV quanta to the chromophores via energy transfer (cf. Figure 5). Perhaps, this last comment best illustrates how difficult it is to answer the question raised in this concluding section.

ACKNOWLEDGMENTS

This work was supported by U.S.P.S. National Institutes of Science grants (NS15426 and GM36956).

REFERENCES

1. **Song, P.-S., Moore, T. A., and Sun, M.,** Excited states of some plant pigments, in *The Chemistry of Plant Pigments,* Chichester, C. O., Ed., Academic Press, New York, 1972, 33.
2. **Song, P.-S. and Moore, T. A.,** On the photoreceptor pigment for phototropism and phototaxis: Is a carotenoid the most likely candidate?, *Photochem. Photobiol.,* 19, 435, 1974.
3. **Song, P.-S.,** Spectroscopic and photochemical characterization of flavoprotein and carotenoproteins as blue light photoreceptors, in *The Blue Light Syndrome,* Senger, H., Ed., Springer-Verlag, Berlin, 1980, 157.
4. **Sarkar, H. K., Song, P.-S., Leong, T. Y., and Briggs, W. R.,** A fluorescence lifetime assay of a membrane-bound flavin from corn coleoptiles, *Photochem. Photobiol.,* 35, 593, 1982.
5. **Presti, D. E.,** The photobiology of carotenes and flavins, in *The Biology of Photoreception,* Cosens, D. J. and Vince-Prue, D., Eds., Cambridge University Press, Cambridge, 1983, 133.
6. **Schmidt, W.,** Artificial flavin/membrane systems; a possible model for physiological blue light action, in *The Blue Light Syndrome,* Senger, H., Ed., Springer-Verlag, Berlin, 1980, 214.
7. **Nakano, K., Sugimoto, T., and Suzuki, H.,** Note on the theory of 3-methyl lumiflavin in solution, *Bull. Sci. Eng. Res. Lab. Waseda Univ.,* 87, 70, 1979.
8. **Nakano, K., Sugimoto, T., and Suzuki, H.,** On the theory of 3-methyl lumiflavin in solution, *J. Phys. Soc. Jpn.,* 48, 939, 1980.
9. **Sun, M., Moore, T. A., and Song, P.-S.,** Molecular luminescence studies of flavins. I. The excited states of flavins, *J. Am. Chem. Soc.,* 94, 1730, 1972.
10. **Grabe, B.,** Electronic structure and spectra of lumiflavin calculated by a restricted Hartree-Fock method, *Acta Chem. Scand.,* 26, 4084, 1972.
11. **Eaton, W. A., Hofrichter, J., Makinen, M. W., Anderson, R. D., and Ludwig, M. L.,** Optical spectra and electronic structure of flavine mononucleotide in flavodoxin crystals, *Biochemistry,* 14, 2146, 1975.
12. **Johansson, L. B. A., Davidson, A., Lindblom, G., and Naqvi, K. R.,** Electronic transitions in the isoalloxazine ring and orientation of flavins in model membranes studied by polarized light spectroscopy, *Biochemistry,* 18, 4249, 1979.
13. **Yu, M. W., Fritchie, C. J., Jr., Fucaloro, A. F., and Anex, B. G.,** Polarization characteristics of flavin spectra. Specular reflectivity of *bis*(10-methylisoalloxazine) copper(II) perchlorate tetrahydrate, *J. Am. Chem. Soc.,* 98,. 6496, 1976.
14. **Sun, M. and Song, P.-S.,** Excited states and reactivity of 5-deazaflavine. Comparative studies with flavine, *Biochemistry,* 12, 4663, 1973.
15. **Schmidt, W.,** Fluorescence properties of isotropically and anisotropically embedded flavins, *Photochem. Photobiol.,* 34, 7, 1981.
16. **Ghetti, F., Colombetti, G., Lenci, F., Campani, E., Polacco, E., and Quaglia, M.,** Fluorescence of *Euglena* photoreceptor pigments: an *in vivo* microspectroscopic study, *Photochem. Photobiol.,* in press.
17. **Smith, E. C. and Metzler, D. E.,** Photochemical degradation of riboflavin, *J. Am. Chem. Soc.,* 85, 3285, 1963.
18. **Moore, W. M., Spence, J. T., Raymond, F. A., and Colson, S. D.,** Photochemistry of riboflavin. I. H transfer process in anaerobic photobleaching of flavins, *J. Am. Chem. Soc.,* 85, 3367, 1963.
19. **Kurtin, W. E., Latino, M. A., and Song, P.-S.,** A study of photochemistry of flavins in pyridine and with a donor, *Photochem. Photobiol.,* 6, 247, 1967.
20. **Song, P.-S. and Metzler, D. E.,** Photochemical degradation of flavins. IV. Studies of the anaerobic photolysis of riboflavin, *Photochem. Photobiol.,* 6, 691, 1967.
21. **Treadwell, G., Cairns, W. L., and Metzler, D. E.,** Photochemical degradation of flavins. V. Chromatographic studies of the products of riboflavin, *J. Chromatogr.,* 35, 376, 1968.
22. **Metzler, D. E. and Cairns, W. L.,** Photochemical degradation of flavins. VI. New photoproduct and its use in studying the photolytic mechanism, *J. Am. Chem. Soc.,* 93, 2772, 1971.
23. **Song, P.-S.,** On the basicity of the excited state of flavins, *Photochem. Photobiol.,* 7, 311, 1968.
24. **Lui, A.,** Photoresponse of *Paramecium* provoked by photosensitizing dyes, *Biol. Glas. Jugosl.,* 8, 11, 1956.

25. **Lozina-Lozinsky, L. K.,** Adaptive behavior of *Parmecium caudatum* to action of photodynamic dyes, *Acta Protozool.,* 18, 609, 1979.

26. **Nultsch, W.,** Photosensing in cyanobacteria, in *Sensory Perception and Transduction in Aneural Organisms,* Colombetti, G., Lenci, F., and Song, P.-S., Eds., Plenum Press, New York, 1985, 147.

27. **Song, P.-S. and Moore, T. A.,** Mechanism of the photodephosphorylation of menadiol diphosphate. A model for bioquantum conversion, *J. Am. Chem. Soc.,* 90, 6507, 1968.

28. **Galston, A. W.,** Riboflavin-sensitized photooxidation of indoleacetic acid and related compounds, *Proc. Natl. Acad. Sci. U.S.A.,* 35, 10, 1949.

29. **Galston, A. W.,** Riboflavin retrospective or deja-vu in blue, *Photochem. Photobiol.,* 25, 503, 1977.

30. **Yagi, K., Ohishi, N., Nishimoto, K., Choi, J. D., and Song, P.-S.,** Effect of hydrogen bonding on electronic spectra and reactivity of flavins, *Biochemistry,* 19, 1553, 1980.

31. **Schmidt, W. and Butler, W. L.,** Flavin-mediated photoreactions in artificial systems, *Photochem. Photobiol.,* 24, 71, 1976.

32. **Song, P.-S.,** Photosensory transduction in *Stentor coeruleus* and related organisms, *Biochim. Biophys. Acta,* 639, 1, 1981.

33. **Yang, S. F., Ku, H. S., and Pratt, H. K.,** Photochemical production of ethylene from methionine and its analogues in the presence of flavin mononucleotide, *J. Biol. Chem.,* 242, 5274, 1967.

34. **Kohler, B.,** Private discussion, *International Conference on Photochemistry and Photobiology,* Alexandria, Egypt, January, 5 to 10, 1983.

35. **Chae, Q., Song, P.-S., Johansen, J. E., and Liaan-Jensen, S.,** Linear dichroic spectra of cross-conjugated carotenals and configurations of in-chain substituted carotenoids, *J. Am. Chem. Soc.,* 99, 5609, 1977.

36. **Schulten, K., Ohmine, I., and Karplus, M.,** Correlation effects in the spectra of polyenes, *J. Chem. Phys.,* 64, 4422, 1976.

37. **Birks, J. B., Tripathi, G. N. R., and Lumb, M. D.,** The fluorescence of all-*trans* diphenyl polyenes, *J. Chem. Phys.,* 33, 185, 1978.

38. **Andrews, J. R. and Hudson, B.,** Geometric effects in the excited states of conjugated trienes, *Chem. Phys. Lett.,* 60, 380, 1979.

39. **Birge, R. R. and Pierce, B. M.,** A theoretical analysis of the two-photon properties of linear polyenes and the visual chromophores, *J. Chem. Phys.,* 70, 165, 1979.

40. **D'Amico, K. L., Manos, C., and Christensen, R. L.,** Electronic energy levels in a homologous series of unsubstituted linear polyenes, *J. Am. Chem. Soc.,* 102, 1777, 1980.

41. **Birge, R. R.,** Photophysics of light transduction in rhodopsin and bacteriorhodopsin, *Annu. Rev. Biophys. Bioeng.,* 10, 315, 1981.

42. **Said, M., Maynau, D., and Malrieu, J.-P.,** Excited-state properties of linear polyenes studied through a nonempirical heisenberg hamiltonian, *J. Am. Chem. Soc.,* 106, 580, 1984.

43. **Thrash, R. J., Fang, H. L. B., and Leroi, F. E.,** The Raman excitation profile spectrum of β-carotene in the preresonance region: evidence for a low-lying singlet state, *J. Chem. Phys.,* 67, 5930, 1977.

44. **Saito, S. and Tasumi, M.,** Resonance raman spectra (5800—40 cm^{-1}) of all-*trans* and 15-*cis* isomers of β-carotene in the solid state and in solutions. Measurements with various laser lines from ultraviolet to red, *J. Raman Spectrosc.,* 14, 299, 1983.

45. **Hudson, B. and Kohler, B. E.,** Linear polyene electronic structure and spectroscopy, *Annu. Rev. Phys. Chem.,* 25, 437, 1974.

46. **Moore, T. A. and Song, P.-S.,** Electronic spectra of carotenoids. β-Carotene, *J. Mol. Spectrosc.,* 52, 209, 1974.

47. **Nebot-Gill, I. and Malrieu, J.-P.,** Biradical or ionic twisted excited states in the singlet *cis-trans* isomerization of polyenes?, *J. Am. Chem. Soc.,* 104, 3320, 1982.

48. **Evstigneev, V. B. and Paramonova, L. I.,** Photoelectrochemical effect in solid films of fucoxanthin, *Biokhimiya,* 41, 548, 1976.

49. **Galland, P. and Lipson, E. D.,** Modified action spectra of photogeotropic equilibrium in *Phycomyces blakesleeanus* mutants with defects in genes madA, madB, madC, and madH, *Photochem. Photobiol.,* 41, 331, 1985.

50. **McDonagh, A. F. M., Palma, L. A., and Lightner, D. A.,** Blue light and bilirubin excretion, *Science,* 208, 145, 1980.

51. **Lamola, A. A., Blumberg, W. E., McCleod, R., and Fanoroff, A.,** Photoisomerized bilirubin in blood from infants receiving phototherapy, *Proc. Natl. Acad. Sci. U.S.A.,* 78, 1882, 1981.

52. **Holzwarth, A. R., Langer, E., Lehner, H., and Schaffner, K.,** Absorption, luminescence, solvent-induced circular dichroism and ^{1}H NMR study of bilirubin dimethyl ester: observation of different forms in solution, *Photochem. Photobiol.,* 32, 17, 1980.

53. **Lamola, A. A., Eisinger, J., Blumberg, W. E., Patel, S. C., and Flores, J.,** Fluorometric study of the partition of bilirubin among blood components: basis for rapid microassays of bilirubin and bilirubin binding capacity in whole blood, *Anal. Biochem.,* 100, 25, 1979.

54. **Briggs, W. R. and Iino, M.,** Blue light-absorbing photoreceptors in plants, *Phil. Trans. R. Soc. London Ser. B.,* 303, 347, 1983.

55. **Margraf, W.,** Orange/yellow pigments in the basidiomycete, *Pleurotus ostreatus* (Jacq. ex. Pr.)Drummer, in *Blue Light Effects in Biological Systems,* Senger, H., Ed., Springer-Verlag, Berlin, 1984, 55.

56. **Thimann, K. V. and Curry, G. M.,** Phototropism, in *Light and Life,* McElroy, W. D. and Glass, B., Eds., Johns Hopkins University Press, Baltimore, 1961, 646.

57. **Diehn, B.,** Action spectra of the phototactic responses in *Euglena, Biochim. Biophys. Acta,* 177, 136, 1969.

58. **Zurzycki, J.,** Blue-light induced intracellular movement, in *The Blue Light Syndrome,* Senger, H., Ed., Springer-Verlag, Berlin, 1980, 58.

59. **Russo, V. E. A.,** Are carotenoids the blue light photoreceptor in the photoinduction of protoperithecia in *Neurospora crassa?, Photochem. Photobiol.,* submitted.

60. **Delbruck, M., Katzir, A., and Presti, D.,** Responses of Phycomyces indicating optical excitation of the lowest triplet state of riboflavin, *Proc. Natl. Acad. Sci. U.S.A.,* 73, 1969, 1976.

61. **Foster, K. W., Saranak, J., Patel, N., Zarilli, G., Okabe, M., Kline, T., and Nakanish, K.,** A rhodopsin is the functional photoreceptor for phototaxis in the unicellular eukaryote *Chlamydomonas, Nature,* 311, 756, 1984.

62. **Majima, T. and Oosawa, F.,** Response of *Chlamydomonas* to temperature change, *J. Protozool.,* 22, 499, 1975.

63. **Hildebrand, E. and Schmiz, A.,** Sensory transduction in *Halobacterium,* in *Sensory Perception and Transduction in Aneural Organisms,* Colombetti, G., Lenci, F., and Song, P.-S., Eds., Plenum Press, New York, 1985, 93.

64. **Spudich, J.,** Color-sensing by phototactic *Halobacterium halobium, Sensory Perception and Transduction in Aneural Organisms,* Colombetti, G., Lenci, F., Song, P.-S., Eds., Plenum Press, New York, 1985, 113.

65. **Vierstra, R. D. and Poff, K. L.,** Role of carotenoids in the phototropic response of corn seedlings, *Plant Physiol.,* 68, 798, 1981.

66. **Foster, K. W. and Smyth, R. D.,** Light antennas in phototactic algae, *Microbiol. Rev.,* 44, 572, 1980.

67. **Shropshire, W., Jr.,** Carotenoids as primary photoreceptors in blue-light responses, in *The Blue Light Syndrome,* Senger, H., Ed., Springer-Verlag, Berlin, 1980, 172.

68. **Schmidt, W.,** The physiology of blue light systems, in *The Biology of Photoreception,* Cosens, D. J. and Vince-Prue, D., Eds., Cambridge University Press, Cambridge, 1983, 306.

69. **Eker, A. P. M., Dekker, R. H., and Berends, W.,** Photoreactivating enzyme from *Streptomyces griseus .* IV. On the nature of the chromophoric cofactor in *Streptomyces griseus* photoreactivating enzyme, *Photochem. Photobiol.,* 33, 65, 1981.

70. **Sancar, A. and Sancar, G. B.,** *Escherichia coli* DNA photolyase is a flavoprotein, *J. Mol. Biol.,* 172, 223, 1984.

71. **Jorns, M. S., Sancar, G. B., and Sancar, A.,** Identification of a neutral flavin radical and characterization of a second chromophore in *Escherichia coli* DNA photolyase, *Biochemistry,* 23, 2673, 1984.

72. **Iwatsuki, M., Joe, C. O., and Werbin, H.,** Evidence that deoxyribonucleic acid photolyase from baker's yeast is a flavoprotein, *Biochemistry,* 19, 1172, 1980.

73. **Horwitz, B. A., Gressel, J., Maldin, S., and Epel, B. L.,** Modified cryptochrome *in vivo* absorption in *Trichoderma dim* photosporulation mutants, *Proc. Natl. Acad. Sci. U.S.A.,* 82, 2736, 1985.

74. **Inoue, Y. and Watanabe, M.,** Perithecial formation in *Gelasinospora reticulispora.* VII. Action spectra in the UV region for the photoinduction and photoinhibition of photoinductive effect brought by blue light, *Plant Cell. Physiol.,* 25, 107, 1984.

75. **Inoue, Y.,** Re-examination of action spectroscopy in blue/near-UV light effects, in *Blue Light Effects in Biological Systems,* Senger, H., Ed., Springer-Verlag, Berlin, 1984, 110.

Chapter 2

PRIMARY REACTIONS AND OPTICAL SPECTROSCOPY OF BLUE LIGHT PHOTORECEPTORS

Werner Schmidt

TABLE OF CONTENTS

I.   Introduction 20

II. Primary Reactions 20
     A.   Relaxation of Excited Physiological Photoreceptor Pigments 20
     B.   Dependence of Conformation on the Flavin Redox State 22
     C.   Effects of Blue Light on Flavoenzymes 23

III. Spectrophotometry of Blue Light Photoreceptors 24
     A.   Turbid Samples 24
     B.   The Muñoz-Butler System (LIAC) 27
     C.   Single-Beam Spectrophotometer 27
     D.   Light-Induced Absorption Changes (LIACs) 30
     E.   Dual-Wavelength Spectrophotometer: the Action Spectrum 30
     F.   Are LIACs Physiologically Relevant? 31
     G.   Artificial Cryptochrome Systems 32

IV. Summary 32

Acknowledgment 33

References 33

I. INTRODUCTION

Today a great variety of physiological blue light responses is known, including directional, metabolic, and photomorphogenetic responses, particularly in lower organisms.[1,3-8] The three main questions — as treated extensively in the present volume — are (1) what is the chemical nature of the blue light receptor (if the action spectrum fulfills specific requirements, the receptor pigment is termed "cryptochrome"; see Senger [9]); (2) where is it localized in the cell; and (3) what are the primary and secondary events that initiate the sensory transduction chain which finally leads to the observed response? A uniform picture for the primary reactions, at least for several blue light responses, appears to crystallize: (1) the photoreceptor pigment is a flavin; (2) it is bound in a highly dichroic manner to a membrane; and (3) the primary reactions are redox in character.

Most likely the term "blue light photoreceptor" does not apply to a single, unique photoreceptor molecule but represents a whole class of physiological blue light-transducing pigment(s). Before the chemical nature of the blue light photoreceptor(s) and its mode of action can be analyzed on a rigorous molecular basis, a clear-cut, unique assay is required. And if the photoreceptor turns out to be a *photochromic* pigment (similar to phytochrome), we would expect two forms either interconvertible by actinic light of different wavelengths or a single blue light-sensitive form, which could be recovered in a dark reversion process following the preceding phototransformation. So far, neither possibility can be excluded.

II. PRIMARY REACTIONS

There are two basic procedures to elucidate sensory pathways: either from the (photo) receptor or from the response end. The present article deals with the first procedure, which depends heavily on specific spectrophotometric apparatuses which often are not commercially available. Following the rule of Grotthus-Draper, sensory transduction is initiated only after electronic excitation (10^{-15} sec) of the receptor pigment. The excited molecule, itself with a lifetime in the picosecond range (singlet state) and physicochemical properties different from those of the ground state, triggers subsequent — mostly unknown — relaxation processes which finally lead to the observed response. By these processes, which are slower than the excitation process by at least six orders, the excitation energy is efficiently conserved in some intermediary products (the first one is most likely the triplet state of the photoreceptor itself).

A. Relaxation of Excited Physiological Photoreceptor Pigments

A sketch of various relaxation processes of electronically excited molecules which have to be considered possible primary reactions of the blue light receptor is given in Figure 1; some are well established in other photoreceptor pigments and easily detected by optical spectroscopy. From the top, the figure shows us the following.

(1) As in chlorophylls, the excited molecule is capable of exchanging electrons with appropriate redox partners. This is a mechanism of primary sensory transduction of blue light physiology favored by most investigators and will be discussed in more detail below ("LIAC"). Instead of electron transfer, the excited molecule might also lose its excitation energy by emission of a light quantum (fluorescence of phosphorescence), a futile process in sensory physiology. However, the complete "trivial" conversion of the excitation energy into heat might well be physiologically relevant ("local heating"),[10] but so far no evidence is available for this. (2) In some cases it is well known that a pK can change by several units upon electronic excitation of a molecule; e.g., the acidobasic properties of the flavin triplet are changed dramatically upon $S_0 \rightarrow S_1 \rightarrow T_1$ transition: T_1 is a base of pyridine strength,[11] while S_0 and S_1 are practically nonbasic (pK = 0). Hence, Hemmerich and

Relaxation processes of photoreceptors

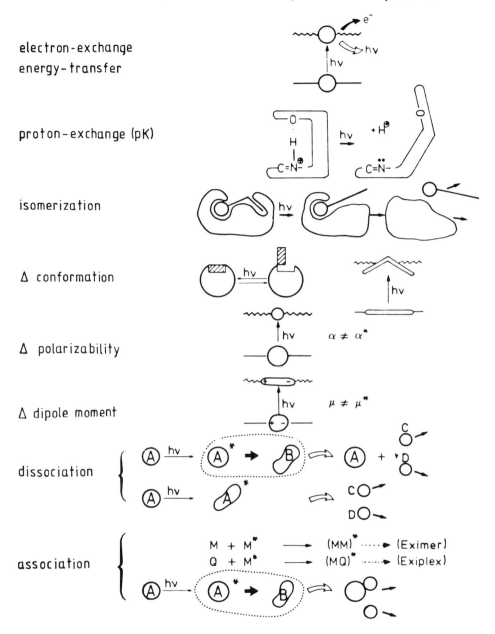

FIGURE 1.   Summarizing sketch of the various relaxation processes of excited molecules (photoreceptors) known, which have to be considered possible primary reactions of the blue light photoreceptor. These include mechanisms well established in other receptors of photobiology, such as proton transfer in bacteriorhodopsin, the 11-cis → all-trans photoisomerization of rhodopsin, conformational changes of phytochrome upon phototransformation P_r → P_{fr}, and others. Spectral changes concomitant with all of these mechanisms are easily detectable by appropriate devices and therefore are widely adopted for analysis. (For details see text.)

Schmidt proposed that a regiospecific rearrangement of hydrogen bridges between flavin and an apoprotein environment may occur in a photoexcited flavoprotein, the blue light receptor, which would induce a unidirectional proton transfer and/or breakage of a hydrogen bridge.[12] (3) The 11-*cis* isomer of retinal as the chromophoric group of rhodopsin is covalently bound via a Schiff's base to the opsin part of the molecule. Upon light absorption the chromophore switches to its all-trans isomer and detaches from the apoprotein. This, in turn, induces a conformational change initiating the process of vision. (4) In the P_r form, the chromophore of the phytochrome molecule protects specific hydrophobic groups, which are exposed upon phototransformation into the P_{fr} form.[13] The other example which also represents this type of relaxation shows conformational changes of the flavin chromophore and is also discussed in some detail below. (5) Changes in polarizability and (6) dipole moments evolve and decay in concert with light absorption; i.e., they are very fast. Their utilization as driving forces for conformational changes which, in turn, could be picked up by the embedding membrane to trigger the effector-response chain is therefore probably very limited. (7) Components C and D as generated by direct or indirect *dissociation reactions* might be capable of initiating the sensory transduction chain. (8) Finally, formation of *eximers* (electronically excited aggregation of two similar molecules) and *exiplexes* (of two different molecules) is widespread among aromatic hydrocarbons.[14] These species have lower-lying excited (singlet) states than the monomers as shown by the bathochromic shift of fluorescence emission spectra compared to the monomers. The dipole moments of exiplexes are typically large (higher than 10 debyes); eximers, on the other hand, have zero dipole moments. We can think of an association of the excited sensitizer pigment molecule with some simultaneously generated dissociation product.[15]

Information continues to accumulate suggesting various light-controlled metabolic pathways, including sensory transduction chains. It is generally believed that flavoenzymes are the proper candidates for these regulations in blue light physiology. Besides the well-known predominant (photo-) redoxproperties of the isoalloxazin- ring and flavoproteins,[16-23] a second feature — which of course is only another aspect of this molecular entity — not so widely appreciated requires our attention in the following paragraph.

B. Dependence of Conformation on the Flavin Redox State

The work of Tauscher et al.[24] has led to the deduction that, in free solution, dihydroflavin ($Fl_{red}H_2$) is in a more-or-less bent configuration, with the degree of bending being enhanced by substituents in position N-5 (Figure 2) as well as by anion formation by deprotonation of position 1. The degree of bending is in keeping with the fact that the central pyrazine ring (eight electrons) is in the planar state, an antiaromatic species of high energy (the pyramidal inversion barrier of the N-5 center is 20 kJ/mol with the degree of bending being directly reflected by the increasing, i.e., more positive, redoxpotential) resisting full planarization; this is particularly true if there is an additional full negative charge in the pyrimidine ring. Thus, the spectrum of the dihydroflavin reflects the degree of bending or puckering of the ring as shown by the long-wavelength transition.[25] The effect of the apoprotein might be either to hinder or enhance this bending. The flavin nucleus in its oxidized state represents a Hückel system, i.e., a resonant, flat molecule. Upon complete photoreduction (i.e., addition of two electrons), the molecule looses aromaticity and attains this so-called "butterfly shape" as sketched in the fourth reaction mode of Figure 1 and shown by the crystallographically obtained structure in Figure 2.[26] Since in most cases investigated a membrane serves as a ternary component, besides the presumed immediate redoxpartner(s), the reaction chain could be triggered by this conformational change as well.

The resonance Raman (RR) and the resonance coherent anti-Stokes Raman scattering (CARS) techniques have been applied recently to free and protein-bound flavins (cf. review by Müller[27]). It may be expected that these techniques will yield information on the interaction

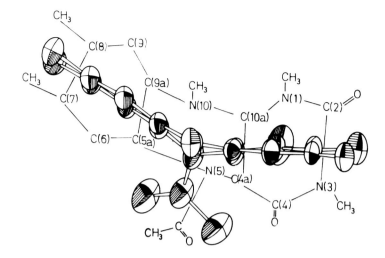

FIGURE 2. Stereoformula of 1,3,10-trimethyl-5-acetyl-1,5-dihydroflavin showing the N5-N10 folding and the noncomplanarity of the 5-acetyl-group (After Kierkkegaard, P., et al., in *Flavins and Flavoproteins*, Kamin, H., Ed., University Park Press, Baltimore, 1971. With permission of University Park Press.)

between the prosthetic group and the flavoprotein, particularly the blue light-photoreceptor. It is now generally accepted that these specific interactions determine the specific biological function of certain flavoproteins.

C. Effects of Blue Light on Flavoenzymes

The influence of light on enzymes in general has been known for a while and was extensively reviewed.[28] Most recently, Ruyters presented the first comprising review with respect to blue light physiology, surveying more than 30 enzymes being influenced by blue light (in most cases *indirectly* via secondary processes; irradiation in vivo[29]). However, only a few (flavo-) enzymes have been shown to be *directly* influenced by blue light — after the addition of *exogeneous* flavin. They are amenable to what Ruyters termed "fine control of enzyme activity by blue light". A change of enzyme conformation is suggested. Symptomatically *(vide infra)*, in most of these cases the enzyme activity is *irreversibly* reduced by blue light;[30,31] only the activity of glycine oxidase[32] and β-D-glucose oxidase[33] are restored in the presence of flavin. However, even an enhancing effect of blue light might reflect some photochemical flavin artifact: glycine oxidase activity was monitored by oxygen consumption, in the presence of as much as 10^{-3} M flavin mononucleotide (FMN).[32] Probably the same arguments apply to the enhancement of activity of glucose oxidase (10^{-3} M FMN, 0.2 mW/cm²). Unfortunately, Schmid did not test the blank (i.e., to monitor the oxygen consumption of the reaction mixture *without* the enzyme present); the present reviewer contends the result would have been the same (cf. Schmid,[32] Figure 2 and Schmidt and Butler,[21] Figure 1). The best-studied example (but still under investigation) is nitrate reductase.[34] This NAD(P)H-dependent molybdo-enzyme of *Neurospora* (and higher plants) comprises both a flavin (flavinadenin nucleotide; FAD) and a b-type cytochrome. The activity of the enzyme is regulated by reversible interconversion between the active (oxidized) and the inactive (reduced) forms. The inactivated enzyme (as accomplished by appropriate reducing agents like dihydroflavin) can be reactivated by blue light, preferentially after addition of 20 μM FAD.[34]

Some general remarks on the photoredox properties of flavin and some criticism on what is termed "conformational change"[4,32,33] are pertinent. Free, solubilized flavins, particularly in the presence of EDTA as a one-electron donor, are highly effective (but fairly unspecific) photoredox-active chromophores.[35] Even in enzymatic concentrations (submicromolar range) they may induce the whole palette of (photo-)chemical reactions,[16,17] particularly under aerobic conditions,[22,36] giving rise to various aggressive oxygen species such as singlet oxygen, hydrogen peroxide, and superoxide (cf. the table in Schmidt[37]). To the flavin photochemist, the order of decreasing photoenzymatic activities of riboflavin (RF), FMN, and FAD[34] appears to reflect trivial photochemical flavin properties[38] rather than physiologically relevant differences (which of course cannot be excluded). The artificial character of flavin photoreactions appears to be manifested if unphysiologically high light intensities (e.g., 130 mW/cm^2, Ninnemann;[34] 24.7 mW/cm^2, Schmid[31]) have to be used for a reaction to occur. Taking typical K_d values of flavoproteins of 10^{-4} into account, the presence of minute amounts of free flavin cannot not be excluded beyond any doubt, not even after extensive dialysis: the statement of photoreduction of nitrate reductase by (internal) bound flavin[34] appears to be questionable. (Mostly inhibiting) effects of blue light on *flavo*enzymes in vitro do not justify the term ''control mechanism'' per se. It is extremely difficult to separate physiologically meaningful from artificial, ''nonsense'' flavin photoreactions. Interpretations of *indirect* effects of blue light on enzymes in vivo (preferentially in nonstarved, i.e., nonstressed, cells) appears to be less questionable.[27,39] However, to the present reviewer it appears to be somewhat conspicuous that ''physiological'' blue light effects are often observed *only* in starved (i.e., somehow stressed) cells (cf. Kowallik and Schänzle[40]), for example, light-induced absorbance changes (LIACs; vide infra) in *Neurospora*[41] or enhanced respiration in *Chlorella*.[42]

Even ''photobinding'' of flavin to protein moieties[32] does not necessarily reflect physiological significance since enhanced and reversible binding of the excited riboflavin in the triplet state to various macromolecules with the effect of an increased lifetime has been known for some 20 years.[43] In any case, they do not provide structural information on the blue light receptor pigment itself. In the strict sense of the molecular biologist, conformational changes — including those of the blue light receptor — can only be verified beyond any doubt by physical means such as X-ray, ORD, CD,[44,45] or ESR analysis. For this purpose the clear-cut characterization and subsequent purification of the receptor pigment is the indispensible requirement. In spite of numerous attempts, research in blue light physiology has not yet reached far beyond the descriptive stadium of phenomenology.

III. BLUE LIGHT PHOTORECEPTORS

A. Turbid Samples

Light absorption by turbid biological materials has been investigated previously in great detail by Shibata,[46]. Butler and Norris,[47] and Butler[48-50] on the basis of Kubelka's theory of intensely scattering material.[51] When dealing with *transparent* materials, the quantity of absorbed light is determined by measuring incident I_0 and transmitted light intensity I, and absorption is defined as A $= -\log T = \epsilon \times d \times c$, with $T = I/I_0$ transmittance, ''ϵ'' extinction coefficient, ''d'' sample thickness, and ''c'' concentration. The logarithm is chosen in order to provide linearity of concentration or sample thickness (solution!) and the detector signal (Beer's law).

When searching for the blue light receptor in vivo we have to expect very low concentrations of the pigment (3×10^9 molecules have been estimated for the growing zone of the *Phycomyces* sporangiophore with a lower absorption limit of A $= 10^{-4}$; Bergman et al.[52]) on top of a highly turbid, scattering material, corresponding to about four absorption units and more, giving rise to absorbancies in units of 10^{-3} (cf. Figure 5; typically, white albino mutants of the various organisms investigated are selected in order to minimize bulk absorption). It is sometimes convenient to think of the absorbancy as composed of three parts: the true absorption, A, the light loss due to scatter, S, and to reflectance, R, corre-

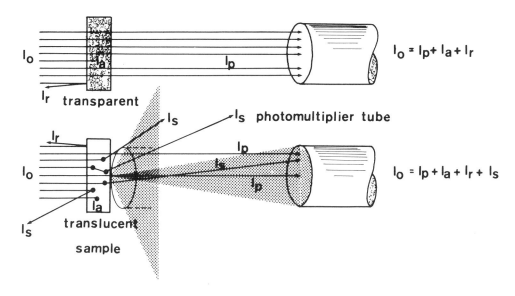

$I_o = I_p + I_a + I_r$

$I_o = I_p + I_a + I_r + I_s$

FIGURE 3.   Comparison of light propagation in transparent (top) and translucent/turbid samples (bottom). *Transparent* samples do not disturb geometrical optics within the spectrophotometer; the exit slit of the monochromator is brought into focus on the cathode of the photomultiplier (PM) tube. Except for the absorbed (I_a) and the reflected light (I_r), all light reaches the PM tube (I_p). The distance between sample and detector is typically several tens of centimeters. However, *translucent/turbid* samples disturb the "light rays" substantially by scattering (I_p). The absorbing centers are inhomogeneously distributed within the sample. Essentially scattered, diffuse light I_s (and very little I_p) remains to be analyzed by the photomultiplier, i.e., light without apparent polarization and little distinguished direction. The most convenient procedure today of collecting as much diffuse light as possible is the usage of end-on PM tubes with photocathodes of large diameter (e.g., 42 mm). The shaded areas indicate the increase of the solid angle, if the PM tube is mounted closer to the photocathode. Compared to common double beam spectrophotometers, such an end-on photomultiplier arrangement allows up to 5000 times higher efficiencies of light collection.

sponding to light intensities I_a, I_s, I_r. Figure 3 demonstrates the light paths different in transparent samples, and in turbid samples relevant in blue light research. We have to remember the fact that the light which is measured in absorption spectroscopy actually has *not* been absorbed by the sample (but strongly scattered in turbid samples). Therefore absorption spectroscopy is an *indirect procedure*, in contrast to *photoacoustic spectroscopy*,[53,54] which detects the actually absorbed quanta (but which so far has not been utilized in blue light physiology). It is easily recognized that a very efficient and versatile device to collect as much light as possible emerging from the sample is the end-on photomultiplier with a large diameter of the photocathode (typically 42 cm), mounted as closely as possible to the sample (Figure 4; see below).

Butler[48] deduced the (simplified) formula for the absorption of a turbid specimen: absorbancy = $0.434b \times S \times d - \log(1 - R^2)$, with $b = 0.5(R^{-1} + R)$. This equation demonstrates the *nonlinear* relationship of sample thickness d and absorbancy.

In a clear solution of a pigment the absorption is defined by $\epsilon \times d \times c$, as mentioned above. Due to multiple scattering within a turbid sample, the actual light path, d, is increased by some factor, β, as great as 20, depending on the specific material giving rise to a β times larger absorption signal.

Since it is unlikely that the cryptochrome molecule will be distributed homogeneously throughout the tissue, the samples can be thought of as suspensions of absorbing particles. Therefore the so-called sieve effect, which has been extensively analyzed by Fukshansky,[55] introduces a further discrepancy between the absorbancies of transparent and translucent samples; that is, different parts of the measuring beam intersect different amounts of cryptochrome (or none; cf. Figure 3).

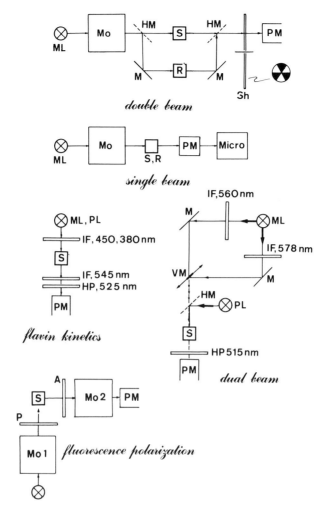

FIGURE 4.   Sketch of the basic spectrophotometric arrangements used in cryptochrome research. Abbreviations used are (A) analyzer; (HP) high-pass filter (blocking filter); (IF) interference filter; (M) mirror; (ML) measuring lamp; (HM) half-mirror, typically 50% transmittance; (Mo) monochromator; (Micro) microcomputer; (P) polarizer; (PL) actinic lamp; (PM) end-on photomultiplier tube; (R) reference cuvette; and (S) sample cuvette. The common *double-beam* spectrometer utilizes a split beam to measure the absorption difference of sample and reference at one wavelength simultaneously; to differentiate between both beams a mechanical chopper (Sh) is required; only transparent samples can be measured. The *single-beam* spectrometer[62] with the PM tube mounted close to the sample and on-line with a microcomputer does not employ any moving part. The optics is simple, allowing the sensitive measurement of true absorption spectra of highly scattering samples. When measuring *kinetics of flavin photoreduction*,[21,22,83] the irradiation lamp serves two purposes: first to provide the strong (actinic) light (PL) required for the photoreduction, and second, to provide the measuring light (ML) exciting fluorescence of the remaining flavoquinone (fully oxidized form of flavin). The *dual-beam* spectrometer determines the (photoinduced) difference in absorbancy between two wavelengths in *one* sample and is the most sensitive spectrophotometric setup available.[66] By means of a blocking filter (HP) and a frequency-selective lock-in amplifier the weak measuring light is separated from the strong actinic light, allowing measurement of flavin-sensitized cyt-b photoreduction during actinic blue light irradiation. For physical reasons, fluorescence spectroscopy is about 1000 times more sensitive than absorption spectroscopy, and flavin concentrations in the nanomolar range are detectable. The capability of the on-line computer in fluorescence spectroscopy, including for the first time continuous polarization-spectra, has been described recently.[65]

Finally, scatter depends on the refractive index differences. In pathing through an absorption band, the refractive index changes with wavelength. As a result, depending upon the direction of scattered radiation, the absorption maximum appears hypsochromically or bathochromically shifted (in the typical sample pigments other than the blue light receptor contribute significant absorption in the blue, e.g., the Soret bands of cytochromes; cf. Figure 5), thereby giving rise to possible additional shifts of 10 and more nanometers, as a simple geometrical consideration shows.

Keeping these arguments in mind, it is clear that optical spectroscopy allows neither the unequivocal characterization and identification of the blue light receptor pigment, nor the determination of its concentration (assuming the molar extincting coefficient of flavins, $\epsilon = 12,400$). Optical spectroscopy as the sole basis for absolute statements remains futile. Spectroscopic data of scattering, turbid biological materials, which often show significant reflectivity of as much as R = 0.8 (cf. Figure 3 and Butler's formula, *vide supra*), yield *qualitative* data, acceptable only to support or to contradict hypotheses derived from non-spectroscopic data.

Besides its spectral properties, the blue light receptor is characterized by additional features (see review by Schmidt[5]): in most instances the primary reaction(s) require(s) oxygen and is/are essentially temperature independent. In contrast to redoxinactive dyes, redoxactive dyes are capable of replacing the natural photoreceptor. Flavin inhibitors are generally also potent inhibitors of blue light physiology. Delbrück and co-workers[56] claimed that they directly induced in *Phycomyces* the quantum-mechanically forbidden singlet-triplet ($S_0 \rightarrow T_1$) transition of flavin in vivo. In contrast to carotenoids, flavins are highly redoxactive, all biologically relevant flavin reactions known are redox in character. This suggests a flavin rather than a carotenoid to serve as blue light receptor pigment, even if the (passive) involvement of carotenoid in physiological blue light action (presumably as a light-guiding pigment similar to its role in photosynthesis) is plausible in several instances.[57,58] Therefore the first process given in Figure 1, a redoxreaction, represents the most likely primary step of the sensory transduction chain in blue light physiology in general.

B. The Muñoz-Butler System (LIAC)

In 1970 Berns and Vaughn[59] demonstrated for the first time blue light-induced spectral changes. Due to technical restrictions at that time, in vivo absorption changes could only be detected at particular wavelengths of the spectrum by dual-wavelength spectroscopy, in contrast to complete spectral changes observed in vitro by fluorimetric procedures. However, Poff and Butler[60] (1974) failed to reproduce their results.

Some years later Muñoz and Butler[41] reported a well reproducible light-induced absorption change (LIAC), which initiated a series of similar experiments (refer to various recent reviews of this topic[1-8]). Because the work of both of these outstanding and prematurely deceased pioneers in photobiology is representative, their experiments on LIACs are adapted to exemplify the procedure and the most relevant optoelectronic equipment utilized in blue light research today. As a blue light-responsive organism they selected the carotene-deficient mutant *poky* of the fungus *Neurospora crassa* (it exhibits various light responses including blue light-induced carotenogenesis and blue light-induced phase shifts of conidiation[61]).

C. Single-Beam Spectrophotometer

The minimum absorption spectrophotometer comprises four entities in a series: a light source, a light selector (monochromator or interference filter), the sample, and a light detector (generally a photomultiplier tube) (Figure 4). The first spectrophotometers were indeed designed on this simple basis. The main disadvantage when monitoring spectra with such a system is a purely technical one: all components exhibit a strong and individual dispersion; i.e., a wavelength-dependent efficiency which heavily truncates the absorption spectra (the

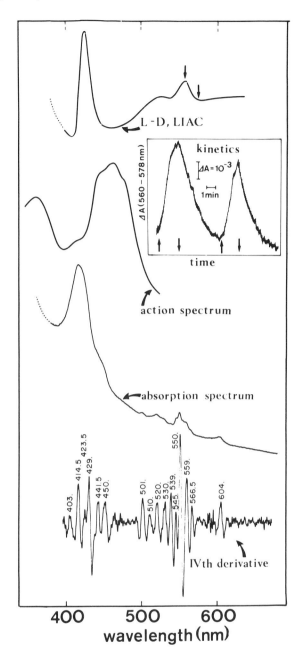

FIGURE 5. L-D, LIAC as measured with the single-beam spectrophotometer at room temperature. Irradiation of mycelium of *Neurospora* induces the dark-reversible photoreduction of a specific b-type cytochrome; the corresponding kinetics as measured in the α-region of the cytochrome with the dual-beam spectrophotometer are shown in the inset. Based on a series of dose-response curves of this LIAC, Muñoz and Butler[41] deduced the action spectrum shown. The absorption spectrum of mycelium mats of *Neurospora* at liquid nitrogen temperature was measured with the single-beam spectrophotometer. This spectrum represents the average of four single scans of the same sample, increasing the SNR by a factor of two. The fourth derivative spectrum was calculated by the computer.[63,64] All peaks marked are well reproducible, reflecting the presence of several pigments obscured in the original spectrum. The L-D spectrum at liquid nitrogen temperature shows a split peak with maxima at 551 and 557 nm.[7] Neither the action spectrum nor the L-D spectrum can be identified in the absorption spectrum in an unequivocal manner.

combined sensitivity of all components is typically maximal in the wavelength range between 450 and 500 nm, a great advantage in blue light receptor research, compared, for example, with phytochrome research, which is confined to the red part of the spectrum). Because of these shortcomings, in the early days of biological spectroscopy the interpretation of spectra was difficult and rendered the proper comparison of spectra as obtained with different spectrophotometers nearly impossible. Additional difficulties were created by the nonlinear dependency of the detector response on the intensity of the impinging light. Finally, temporal fluctuations of light source and sample gave rise to noisy and badly reproducible spectra and kinetics. All these problems were largely overcome by the ingenious invention of the double-beam spectrophotometer, essentially adopting a *compensatory principle*, in time, space, and sensitivity. However, in order to collect a sufficient amount of measuring light on the photocathode of the photomultiplier, the indispensible requirement for those measurements is the transparency of the sample. Turbid materials, particularly those investigated in blue light physiology, have to be excluded (Figure 3).

But modern materials of photomultiplier cathodes and elaborate electronic circuitry such as analog/digital and log converters, and a highly stabilized power supply for the light source improved the signal stability and the signal to noise ratio (SNR) significantly. Therefore, with the feasiblity of running a modestly priced personal computer on line with the photomultiplier tube, the former ingenious double-beam arrangement of absorption spectrophotometers not only became meaningless, but now it even interferes with the computer mode of absorption spectroscopy (i.e., frequency of data acquisition and lightchopper frequency). This called for a rebirth of the plain single-beam spectrophotometer with simple optics and without moving parts like lightchoppers (which, in turn, give rise to another type of noise). Using a single beam spectrophotometer, the "spectra" of reference ($\log I_0 = f(\lambda)$) and sample ($\log I = f(\lambda)$) are monitored *sequentially* rather than simultaneously, and the difference $\log I_0 - \log I$ is calculated (and plotted) by the computer.

In principle even the logarithmic amplifier (which converts transmission into absorbance electronically) is superfluous and could be replaced by computer software calculating the logarithm of the photomultiplier signal; however, such a rigorous procedure is not yet followed routinely, since the real absorption spectra are even harder to visualize on the basis of uncorrected transmission than absorption spectra while scanning. Since in the single-beam setup the light path for sample and reference are virtually identical, the sensitivity of absorption measurements is dramatically improved compared to common double-beam spectroscopy (by a factor of 10 and more; i.e., the "baseline" remains flat at 10 times higher sensitivities). Compared to conventional absorption spectrophotometers, the single-beam setup allows a very close distance between sample and photomultiplier cathode (less than 1 cm) with large diameter, increasing the room angle for light collection by a factor of approximately 5000 (Figure 3). Splitting the optical light path into two portions indispensibly requires a large distance between sample and photomultiplier tube (tens of centimeters) which, however, is barely acceptable for the measurement of turbid materials (this problem was partly overcome by the classical "opal glass method" by Shibata[46]).

Absorption spectra (e.g., of *Neurospora*) are assayed as follows.[41] Mycelium is gently pressed onto the bottom of a steel cuvette. For enhanced wavelength resolution, the absorption spectrum is monitored at liquid nitrogen temperature ($-196°C$) (for details see Schmidt[62]). A reference mimicking the scattering properties of the sample is prepared from several layers of white, nonfluorescent tissue paper. The number of layers is adjusted to yield the same photomultiplier current (e.g., $10^{-5}A$; EMI, type 9608, S20) at a wavelength where the sample does not contribute any significant absorption (700 nm in *Neurospora*, cf. Figure 5). Subtraction of the reference "spectrum" from the (uncorrected) absorption spectrum by the computer immediately yields the corrected absorption spectrum as shown in Figure 5 (another, more elegant procedure which avoids tissue paper as reference but utilizes the sample material itself was described by Schmidt[62] and yields the same result).

The fourth derivative spectrum (Figure 5, bottom) as calculated by the computer according to the procedure of Butler and Hopkins[63,64] with an (average) resolution of dx = 2 nm exhibits a great number of components in a well reproducible manner, particularly cytochromes a, b, and c. As explained for the absorption spectrum, the fourth derivative spectrum should only be accepted if the difference of fourth derivatives deduced from different scans of the same spectrum yields a horizontal line.

D. Light-Induced Absorption Changes (LIACs)

On the assumption that the blue light photoreceptor and/or one of its immediate reaction partners might change in response to physiologically active light, photoinducible absorbance changes (LIACs) were sought in *Neurospora* mycelium on a time scale from seconds to minutes. In their original work, Muñoz and Butler[41] used a monochromator capable of scanning the whole visible spectrum within about 5 sec. They prepared a sample as described before. While aging (i.e., starving and going anaerobic) in the cuvettes, cytochromes and flavin(s) slowly oxidize and become increasingly amenable to a dark-reversible photoreduction by blue light. One spectrum (uncorrected for the apparatus response) was monitored in darkness, D (i.e., without actinic light irradiation). Then the sample was irradiated for some 60 sec with blue light of approximately 1 mW/cm^2 (within the sample compartment, without touching it to guarantee identical optical geometry; *vide supra*); immediately after this irradiation the spectrum was monitored, L. The amplified (and smoothed [Schmidt[65]]) difference spectrum (Figure 5, L-D) solely reflects the photodynamic action of light on the sample. The apparatus response drops out. Clearly, blue light induces the photoreduction of one specific b-type cytochrome.

Another procedure to monitor LIACs is by low-temperature spectroscopy. Two samples of mycelium as similar as possible in their optical properties are prepared. One sample is frozen to liquid nitrogen temperature ($-196°C$) in the dark, and the other under blue light irradiation; both spectra are measured (up to eight times each and accumulated, *vide supra*) and stored in the computer memory. Again, the amplified difference (L-D) reveals the photoreduction of a b-type cytochrome. The advantage of the low-temperature procedure is the increased wavelength resolution (Figure 5). The disadvantage is that only a small wavelength range can be covered (typically the α range; cf. Figure 6 in Schmidt[7]). It is not feasible to prepare two samples with identical optical properties throughout the whole spectrum covered by cytochromes; and it is impossible to use only a single individual sample for both spectra (D and L), since repeated freezing/thawing cycles change both its physiological and optical properties.

E. Dual-Wavelength Spectrophotometer: the Action Spectrum

Once the LIAC-spectrum is known, its kinetic behavior is attainable by dual-wavelength spectroscopy (Figure 4). This procedure makes it possible to monitor very small absorption changes between two wavelengths (Figure 5) defined by light beams of low intensity (10^2 erg/cm^2 sec), as induced by simultaneously impinging much stronger actinic light (10^5 erg/cm^2 sec.[66]). The background absorption might be as high as four to five absorption units. The (difference) signal is stable and the SNR extraordinary low. Depending on the time constant of the lock-in amplifier used, absorption changes smaller than 10^{-4} are detectable[66]). These features are obtained by "clamping" the observed signal changing at λ_1 to a signal from another point in the spectrum λ_2 of the *same* sample, where no change takes place. This compensates for fluctuations of the measuring lamp, the sample, and the photomultiplier voltage. In spite of modern circuitry and stable light sources, single-beam machines cannot compete with the "dual-beam spectrometer" regarding sensitivity.

The inset of Figure 5 shows the kinetics of the dark-reversible cytochrome-b (cyt-b) photoreduction, monitored at the -peak at 560 nm and "fixed" to 578 nm, where no LIAC takes place. On this basis Muñoz and Butler[41] deduced the "true" action spectrum (for the theory of action spectroscopy see Shropshire[67]) for the LIAC in *Neurospora* (reproduced in Figure 5) from a series of dose-response curves. Clearly, cyt-b itself is *not* the photoreceptor pigment, but is photoreduced by a flavin moiety. The gross structure of this action spectrum closely resembles the blue light action spectra of *Neurospora*. According to the operational definition,[9] this photoreceptor is *cryptochrome*. Therefore, the kinetics of the LIAC are taken to reflect some mechanism(s) involved in the primary reactions of blue light physiology (discussion below). During the 10 years following the discovery by Muñoz and Butler[41] various LIACs have been described and analyzed. However, up until now LIACs did not work out convincingly as a unique, unequivocal assay for cryptochrome, in analogy to the photoreversibility of phytochrome (however, even the "clear-cut" reversibility of phyto-chrome might give rise to some pitfalls[68]).

F. Are LIACs Physiologically Relevant?

There are several arguments in favor of a negative response to this question. First, most organisms which are physiologically sensitive to blue light — e.g., etiolated and highly phototropic corn coleoptiles — do *not* show any specific LIAC, when *freshly* prepared. Second, unlike phytochrome, the blue light receptor coincides spectrally with many other pigments like flavins and cytochromes obligatory present in all cells (cf. Figure 5). Third, strong actinic light, as typically required in these experiments (range 1 to 100 mW/cm^2), always induces some kind of *unspecific* LIAC (including trivial bleaching), involving various pigments present in any tissue. Typical intensities of actinic blue light used in LIAC ex-periments *(vide supra)* exceed the physiologically required intensities by more than five orders. Fourth, flavins, which are ubiquitous components of all cells, generally in protein-bound form, are among the most potent sensitizers known in photochemistry, especially in the *free* form of the chromophore *(vide supra)*. Fifth, HeLa cells, which do not show any physiological response to blue light, exhibit "typical" LIACs in the blue.[69] Sixth, the quantum efficiency for electron transport reflected by a LIAC is generally too low to be physiologically relevant (LIACs in pellets of corn and *Neurospora* with quantum efficiencies near unity are the known exceptions[70]). Seventh, in addition to the dark-reversible photo-reduction of a b-type cytochrome, nearly all LIACs reported so far, particularly those in vivo, exhibit a photoreduction rather than the expected photooxidation of a flavin-type pigment (cf. Figure 5); the concentration of these flavin(s) often exceeds the concentration of photoreduced cyt-b by a factor of more than ten. Other LIACs indicate the concomitant photoreduction of cyt-a in addition to cyt-b,[71] and even of cyt-c.[21] This questions a linear electron transport chain, rather suggesting a general photoinduced decrease of the redox potential as mediated by flavin. Another remark is pertinent: cytochrome-c is hydrophilic and easily lost during any preparation; even in rather crude pellets, b- and a-type cytochromes remain,[21] not providing a test for "purity for plasamembranes".

In contrast to these negative arguments, there also exist several lines of evidence in support of the physiological relevance of LIACs. Ulaszewski and co-workers[72] reported that various yeast mutants containing b-type cytochromes are inhibited by the high irradiance of blue light (like the wild type), whereas mutants lacking b cytochromes are not affected. Also, results reported by Klemm and Ninnemann[73] and Ninnemann and Klemm-Wolfgramm[71] show a correlation between a LIAC and a light-dependent physiological response in *Neu-rospora* (light-induced conidiation), probably mediated by nitrate reductase *(vide supra)*. However, this interpretation was questioned by Paietta and Sargent,[74] in view of their finding that photoresponses assayed in nitrate-reductase-deficient mutants showed *no* significant difference as compared to responses in a strain that could utilize nitrate. But the mutant *poky* of *Neurospora* which is deficient in all its b-type cytochromes, exhibits only about 1% LIAC of the wild type, and is less than 1% as physiologically sensitive as the wild type.[75]

Consistently, Paietta and Sargent[76] observed a parallel effect in riboflavin mutants of *Neurospora*, in which a reduction in light sensitivity correlated with flavin deficiency.

As an example for higher plants, Goldsmith et al.[77] have described photoreactions in membrane fractions prepared from maize coleoptiles. Leong et al.[78] have purified and characterized from maize coleoptiles a cytochrome/flavin complex responsible for the LIAC. They presented evidence that the herbicide acifluorofen is acting specifically on this complex which most likely serves to transduce the light signal into curvature in phototropism in oats, with the flavin moiety itself being the photoreceptor.[79] Moreover, inhibitors that are effective in blocking the photoreduction in vivo[80] preferentially inhibited the phototropic over the geotropic response of corn coleoptiles.[81] But again, the physiologic relevance of these findings with respect to phototropism of grass seedlings (and probably all other nonstarved and nontreated tissues) remains questionable since specific and dark-reversible LIACs have never been observed in vivo.

G. Artificial Cryptochrome Systems

Optical spectroscopy of (artificial) membranes has been reviewed in detail previously.[82] Due to limited space, only some accomplishments of optical spectroscopy of artificial flavin (cryptochrome)/membrane systems are listed here. It is reasonable to assume that the relaxation of the blue light receptor is immediately succeeded by a redox reaction *(vide supra),* which — by analogy to better understood primary reactions in photobiology — induces a proton, and/or an electrical, and/or a redox gradient across a membrane. Those reactions appear to be physiologically highly efficient primary sensory transduction steps generally adopted by nature.

On this basis we performed various spectroscopic approaches utilizing a model system, designed to mimic the natural blue light receptor system, including the membrane side. As a minimum molecular model, we have therefore synthesized three different amphiphilic flavins bearing C_{18}-hydrocarbon chains at various positions on the flavin nucleus and anchored them within artificial single-shelled vesicles made from various phospholipids[37,83-86] ("anisotropic flavin chemistry", in contrast to the well-investigated "isotropic flavin chemistry"[16,17]). Due to the diameter of the vesicles of 25 to 30 nm, these suspensions are highly scattering, a condition which, again, requires specific optospectroscopic equipment for the measurement of turbid materials described above (Figure 4). In fact, scattering properties themselves can be adopted for the chromatographical purification of vesicles and to monitor lipid phase transitions from the *gel-* to the *liquid-crystalline* phase.[86] Principally, conformational changes of the blue light receptor pigment could also be detected by this sensitive, noninvasive procedure.

Using various types of photospectroscopic procedures such as are sketched in Figure 4 (discussed above except fluorimetric procedures; see the legend to Figure 4), we have found that virtually all flavin properties are strongly and specifically modified under membrane-bound conditions by lipid type, lipid phase, and by the specific orientation and localization of the flavin nucleus within the membrane. The membrane controls both the thermodynamic and kinetic properties, and the transition dipoles of flavin, the flavin "self-contact" (the predominant and highly disturbing flavin reaction in solution), is suppressed.[85] Singlet *and* triplet states are involved in anisotropic flavin chemistry.[86] Blue light-induced, flavin-mediated membrane transport of redox equivalents and protons is studied and turns out to be highly complex; the involvement of superoxide, flavosemiquinone, dihydroflavin, singlet oxygen, and hydrogen peroxide is suggested.[37]

IV. SUMMARY

The investigation of the primary reactions of cryptochrome, the presumed physiological blue light photoreceptor(s) mediating numerous biological responses, depends intimately on proper optospectroscopic procedures. This paper elucidates some of the threatening problems and pitfalls, which the investigator in this area has to master or has to be aware of. Various optospectroscopic devices, such as are utilized in blue light physiology are presented.

ACKNOWLEDGMENT

I wish to thank the late Prof. Dr. W. L. Butler for the invaluable introduction into optical spectroscopy of biological photoreceptors and all kinds of related techniques which cannot be learned from textbooks.

REFERENCES

1. **Senger, H., Ed.,** *The Blue Light Syndrome,* Springer-Verlag, Berlin, 1980.
2. **Senger, H.,** The effect of blue light on plants and microorganisms, *Photochem. Photobiol.,* 35, 911, 1982.
3. **Senger, H., Ed.,** *Blue Light Effects in Biological Systems,* Springer-Verlag, Berlin, 1984.
4. **Senger, H. and Briggs, W. R.,** The blue light receptor(s): primary reactions and subsequent metabolic changes, in *Photochemical and Photobiological Reviews,* Vol. 6, Smith, K. C., Ed., Plenum Press, New York, 1981, chap. 1.
5. **Schmidt, W.,** Physiological bluelight reception, in *Structure and Bonding,* Vol. 41, *Molecular Structure and Sensory* Transduction, Hemmerich, P., Ed., Springer-Verlag, Berlin, 1980, chap. 1.
6. **Schmidt, W.,** The physiology of blue-light systems, in *The Biology of Photoreception,* Cosens, D. J. and Vince-Prue, D., Eds., Cambridge University Press, Cambridge, 1983, 305.
7. **Schmidt, W.,** Bluelight physiology, *BioScience,* 34, 698, 1984.
8. **Briggs, W. R. and Iino, M.,** Blue-light-absorbing photoreceptors in plants, *Phil. Trans. R. Soc. London Ser. B,* 303, 347, 1983.
9. **Senger, H.,** Chryptochrome, some terminological thoughts, in *Blue Light Effects in Biological Systems,* Senger, H., Ed., Springer-Verlag, Berlin, 1984, 72.
10. **Song, P.-S.,** Spectroscopic and photochemical characterization of flavoproteins and carotenoproteins as blue light photoreceptors, in *The Blue Light Syndrome,* Senger, H., Ed., Springer-Verlag, Berlin, 1980, 158.
11. **Schreiner, S., Steiner, U., and Kramer, H. E. A.,** Determination of the pK values of the lumiflavin triplet state by flash photolysis, *Photochem. Photobiol.,* 21, 81, 1975.
12. **Hemmerich, P. and Schmidt, W.,** Blue light reception and flavin photochemistry, in *Photoreception and Sensory Transduction in Aneural Organisms,* Lenci, F. and Colombetti, G., Eds., Plenum Press, New York, 1980, 271.
13. **Song, P.-S.,** The molecular basis of phytochrome (Pfr) and its interactions with model receptors, in *The Biology of Photoreception,* Cosens, D. J. and Vince-Prue, D., Eds., Cambridge University Press, Cambridge, 1983, 181.
14. **Turro, N. J.,** *Molecular Photochemistry,* W. A. Benjamin, New York, 1965, chap. 7.
15. **Barltrop, J. A. and Coyle, J. D.,** *Principles of Photochemistry,* John Wiley & Sons, Chichester, England, 1978, chap. 4.
16. **Bruice, T. C.,** Models and flavin catalysis, in *Progress in Bioorganic Chemistry,* Vol. 4, Kaiser, E. T. and Kezdy, F. J., Eds., John Wiley & Sons, New York, 1976, 1.
17. **Hemmerich, P.,** The present status of flavin and flavocoenzyme chemistry, in *Fortschritte der Chemie organischer Naturstoffe,* Herz, W., Grisebach, H., and Kirby, G. W., Eds., Springer-Verlag, Berlin, 1976, 451.
18. **Haas, W. and Hemmerich, P.,** Flavin dependent substrate photooxidation as a chemical model of dehydrogenase action, *Biochem. J.,* 181, 95, 1979.
19. **Massey, V. and Hemmerich, P.,** Active-site probes of flavoproteins, *Biochem. Soc. Trans. (Biochem. Rev.,)* 8, 246, 1980.
20. **Massey, V., Stankovich, M., and Hemmerich, P.,** Light-mediated reduction of flavoproteins with flavins as catalysts, *Biochemistry,* 17, 1, 1978.
21. **Schmidt, W. and Butler, W. L.,** Light-induced absorbance changes in cell-free extracts of *Neurospora crassa, Photochem. Photobiol.,* 24, 77, 1976.

22. **Schmidt, W. and Butler, W. L.,** Flavin-mediated photoreactions in artificial systems: a possible model for the blue light photoreceptor pigment in living systems, *Photochem. Photobiol.*, 24, 71, 1976.

23. **De Kok, A., Veeger, C., and Hemmerich, P.,** The effect of light on flavins and flavoproteins in the presence of α-keto acids, in *Flavins and Flavoproteins*, Kamin, H., Ed., University Park Press, Baltimore, 1971, 63.

24. **Tauscher, L., Ghisla, S., and Hemmerich, P.,** NMR study of nitrogen inversion of conformation of 1,5-dihydro-isoalloxazines ("reduced flavin"), *Helv. Chim. Acta*, 56, 630, 1973.

25. **Dudley, K. H., Ehrenberg, A., Hemmerich, P., and Müller, F.,** Spektren und Strukturen der am Flavin-Redoxsystem beteiligten Partikeln. Studien in der Flavinreihe. IX, *Helv. Chim. Acta*, 47, 1354, 1964.

26. **Kierkkegaard, P., Norrestam, R., Werner, P., Csöregh, I., Glehn, M., Karlsson, R., Leijonmarck, M., Rönnquist, O., Stennesland, B., Tillberg, O., and Torbjörnsson, T.,** X-ray structure investigation of flavin derivatives, in *Flavins and Flavoproteins*, Kamin, H., Ed., University Park Press, Baltimore, 1971, 1.

27. **Müller, F.,** Spectroscopy and photochemistry of flavins and flavoproteins, *Photochem. Photobiol.*, 34, 753, 1981.

28. **Anderson, L. E.,** Interaction between photochemistry and activity of enzymes, in *Encyclopedia of Plant Physiology*, New Series, Vol. 6, Gibbs, M. and Latzko, E., Eds., Springer-Verlag, Berlin, 1979, 271.

29. **Ruyters, G.,** Effects of blue light on enzymes, in *Blue Light Effects in Biological Systems*, Senger, H., Ed., Springer-Verlag, Berlin, 1984, 283.

30. **Codd, G. A.,** The photoinactivation of tobacco transketolase in the presence of flavin mononucleotide, *Z. Naturforsch.*, 27b, 701, 1972.

31. **Schmid, G. H.,** The effect of blue light on some flavin enzymes, *Hoppe-Seyler's Z. Physiol. Chem.*, 351, 575, 1970.

32. **Schmid, G. H.,** Conformational changes caused by blue light, in *The Blue Light Syndrome*, Senger, H., Ed., Springer-Verlag, Berlin, 1980, 198.

33. **Schmid, G. H.,** Photoregulation of β-D-glucose oxidase by blue light, *Photochemistry*, 10, 2041, 1971.

34. **Ninnemann, H.,** The nitrate reductase systems, in *Blue Light Effects in Biological Systems*, Senger, H., Ed., Springer-Verlag, Berlin, 1984, 95.

35. **Knappe, W.-R.,** Dihydro derivatives of flavins substituted with electron withdrawing groups, in *Flavins and Flavoproteins*, Yagi, K. and Yamano, T., Eds., University Park Press, Baltimore, 1980, 469.

36. **Penzer, G. R.,** The chemistry of flavins and flavoproteins: aerobic photochemistry, *Biochem. J.*, 116, 733, 1970.

37. **Schmidt, W.,** Blue light-induced, flavin-mediated transport of redoxequivalents across artificial bilayer membranes, *J. Membr. Biol.*, 82, 113, 1984.

38. **Yagi, K.,** Free and bound flavin fluorescence, in *Biochemical Fluorescence: Concepts*, Vol. 2, Chen, R. F. and Edelhoch, H., Eds., Marcel Dekker, New York, 1976, chap. 15.

39. **Andersag, R. and Pirson, A.,** Verwertung von glukose in chlorellakulturen bei Blau- und Rotlichtbestrahlung, *Biochem. Physiol. Pflanz.*, 169, 71, 1976.

40. **Kowallik, W. and Schätzle, S.,** Enhancement of carbohydrate degradation by blue light, in *Light Effects in Biological Systems*, Senger, H., Ed., Springer-Verlag, Berlin, 1984, 344.

41. **Muñoz, V. and Butler, W. L.,** Photoreceptor pigment for blue light in *Neurospora crassa*, *Plant Physiol.*, 55, 421, 1975.

42. **Kowallik, W. and Gaffron, H.,** Respiration induced by blue light, *Planta*, 69, 92, 1966.

43. **Kostenbauer, H. B., Deluca, P. P., and Kowarski, C. R.,** Photobinding and photoreactivity of riboflavin in the presence of macromolecules, *J. Pharm. Sci.*, 54, 1243, 1965.

44. **Scola-Nagelschneider, G. and Hemmerich, P.,** Circular dichroism, self interaction and side chain conformation of riboflavin and riboflavin analogues, *Z. Naturforsch.*, 27b, 1044, 1972.

45. **Eweg, J. K., Müller, F., and van Berkel, W. J. H.,** On the enigma of old yellow enzyme's spectral properties, *J. Biochem.*, 129, 303, 1982.

46. **Shibata, K.,** Spectrophotometry of translucent biological materials — opal glass transmission method, in *Methods of Biochemical Analysis*, Vol. 7, Glick, D., Ed., Interscience, New York, 1957, 77.

47. **Butler, W. L. and Norris, K. H.,** The spectrophotometry of dense light-scattering material, *Arch. Biochem. Biophys.*, 87, 31, 1960.

48. **Butler, W. L.,** Absorption of light by turbid materials, *J. Opt. Soc. Am.*, 52, 292, 1962.

49. **Butler, W. L.,** Absorption spectroscopy *in vivo*: theory and application, *Annu. Rev. Plant Physiol.*, 15, 1964, 451.

50. **Butler, W. L.,** Absorption spectroscopy of biological materials, in *Methods in Enzymology*, San Pietro, A., Ed., Academic Press, New York, 1972, 3.

51. **Kubelka, P.,** New contributions to the optics of intensively light-scattering material, *J. Opt. Soc. Am.*, 38, 1948, 448.

52. **Bergman, K., Burke, P. V., Cerdá-Olmedo, E., David, C. N., Delbrück, M., Foster, K. W., Goodell, E. W., Heisenberg, M., Meissner, G., Zalokar, M., Dennison, D. S., and Shrophire, W., Jr.,** Phycomyces, *Bacteriol. Rev.,* 33, 99, 1969.

53. **Moore, T. A.,** Photoacoustic spectroscopy and related techniques applied to biological materials, in *Photochemical and Photobiological Reviews,* Vol. 7, Smith, K. C., Ed., Plenum Press, New York, 1983, chap. 4.

54. **Betteridge, D. and Meylor, P. J.,** Analytical aspects of photoacoustic spectroscopy, *Crit. Rev. Anal. Chem.,* 14, 267, 1984.

55. **Fukshansky, L.,** On the theory of light absorption in non-homogeneous objects, *J. Math. Biol.,* 6, 177, 1978.

56. **Delbrück, M., Katzir, A., and Presti, D.,** Response of *Phycomyces* indicating optical excitation of the lowest triplet state of riboflavin, *Proc. Natl. Acad. Sci., U.S.A.,* 73, 1969, 1976.

57. **Shropshire, W. Jr.,** Carotenoids as primary photoreceptors in blue light responses, in *The Blue Light Syndrome,* Senger, H., Ed., Springer-Verlag, Berlin, 1980, 172.

58. **DeFabo, E. C., Harding, R. W., and Shropshire, W., Jr.,** Action spectrum between 260 and 800 nanometers for the photoinduction of carotenoid biosynthesis in *Neurospora crassa, Plant Physiol.,* 57, 440, 1976.

59. **Berns, D. S. and Vaughn, R.,** Studies on the photopigment system in *Phycomyces, Biochem. Biophys. Res. Commun.,* 39, 1094, 1970.

60. **Poff, K. L. and Butler, W. L.,** Absorbance changes induced by blue light in *Phycomyces blakesleeanus* and *Dictyostelium discoideum, Nature (London),* 248, 799, 1974.

61. **Feldman, J. F. and Dunlap, J. C.,** *Neurospora crassa,* a unique system for studying circadian rhythms, in *Photochemical and Photobiological Reviews,* Vol. 7, Smith, K. C., Ed., Plenum Press, New York, 1983, chap. 7.

62. **Schmidt, W.,** A computerized single-beam spectrophotometer: an easy setup, *Anal. Biochem.,* 125, 162, 1982.

63. **Butler, W. L. and Hopkins, D. W.,** Higher derivative analysis of complex absorption spectra, *Photochem. Photobiol.,* 12, 439, 1970.

64. **Butler, W. L. and Hopkins, D. W.,** An analysis of fourth derivative spectra, *Photochem. Photobiol.,* 12, 451, 1970.

65. **Schmidt, W.,** On-line computer capability in fluorescence spectroscopy, *Opt. Eng.,* 22, 576, 1983.

66. **Schmidt, W.,** A high performance dual-wavelength spectrophotometer and fluorometer, *J. Biochem. Biophys. Meth.,* 2, 171, 1980.

67. **Shropshire, W., Jr.,** Action spectroscopy, in *Phytochrome,* Mitrakos, K. and Shrophsire, J. R., Eds., Academic Press, New York, 1972, 162.

68. **Schmidt, W.,** Red/far-red photoreversibility: not an appropriate phytochrome assay for red-light preirradiated corn coleoptiles, *Photochem. Photobiol.,* 39, 267, 1984.

69. **Lipson, E. and Presti, D.,** Light-induced absorbance changes in *Phycomyces* photomutants, *Photochem. Photobiol.,* 25, 203, 1977.

70. **Lipson, E. and Presti, D.,** Graphical estimation of cross sections from fluence response data, *Photochem. Photobiol.,* 32, 383, 1980.

71. **Ninnemann, H. and Klemm-Wolfgramm, E.,** Blue light-controlled conidiation and absorbance change in *Neurospora* are mediated by nitrate reductase, in *The Blue Light Syndrome,* Senger, H., Ed., Springer-Verlag, Berlin, 1980, 238.

72. **Ulascewski, S., Mamounas, T., Shen, W.-R., Rosenthal, P. J., Woodward, J. R., and Edmunds, L. N., Jr.,** Light effects in yeast: evidence for participation of cytochromes in photoinhibition of growth and transport in *Saccharomyces cervisiae* cultured at low temperatures, *J. Bacteriol.,* 138, 523, 1979.

73. **Klemm, E. and Ninnemann, H.,** Correlation between absorbance changes and a physiological response induced by blue light in *Neurospora, Photochem. Photobiol.,* 28, 227, 1978.

74. **Paietta, J. and Sargent, M. L.,** Blue light responses in nitrate redutase mutants of *Neurospora crassa, Photochem. Photobiol.,* 35, 853, 1982.

75. **Brain, R. D., Woodward, D. O., and Briggs, W. R.,** Correlative studies of light sensitivity and cytochrome content in *Neurospora crassa, Carnegie Inst. Washington Yearb.,* 76, 295, 1977b.

76. **Paietta, J. and Sargent, M. L.,** Photoreception in *Neurospora crassa:* correlation of reduced light sensitivity with flavin deficiency, *Proc. Natl. Acad. Sci. U.S.A.,* 78, 5573, 1981.

77. **Goldsmith, M. H. M., Caubergs, R. J., and Briggs, W. R.,** Light-inducible cytochrome reduction in membrane preparations from corn coleoptiles, *Plant Physiol.,* 66, 1067, 1980.

78. **Leong, T.-Y., Vierstra, R. D., Briggs, W. R.,** A blue light-sensitive cytochrome-flavin complex from corn coleoptiles. Further characterization, *Photochem. Photobiol.,* 34, 697, 1981.

79. **Leong, T.-Y. and Briggs, W. R.,** Evidence from studies with actifluorfen for the participation of a flavin-cytochrome complex in blue light photoreception for phototropism of oat coleoptiles, *Plant Physiol.,* 70, 875, 1982.

80. **Caubergs, R. J., Goldsmith, M. H. M., and Briggs, W. R.,** Light-inducible cytochrome reduction in membranes from coleoptiles: fractionation and inhibitor studies, *Carnegie Inst. Washington Yearb.,* 78, 121, 1979.

81. **Schmidt, W., Hart, J., Filner, P., and Poff, K. L.,** Specific inhibition of phototropism in corn seedlings, *Plant Physiol.,* 60, 736, 1977.

82. **Heesemann, J. and Zingsheim, H. P.,** Optical spectroscopy of monolayers, multilayer assemblies, and single model membranes, in *Molecular Biology, Biochemistry,* Grell, E., Ed., Springer-Verlag, Berlin, 172, 1981.

83. **Schmidt, W.,** On the environment and rotational motion of amphiphilic flavins in artificial membrane vesicles as studied by fluorescence, *J. Membr. Biol.,* 47, 1, 1979.

84. **Schmidt, W. and Hemmerich, P.,** On the redox reactions and accessibility of amphiphilic flavins in artificial membrane vesicles, *J. Membr. Biol.,* 60, 129, 1981.

85. **Schmidt, W.,** Fluorescence properties of isotropically and anisotropically embedded flavins, *Photochem. Photobiol.,* 34, 7, 1981.

86. **Schmidt, W.,** Further photophysical and photochemical characterization of flavins associated with single-shelled vesicles, *J. Membr. Biol.,* 76, 73, 1983.

Chapter 3

ACTION SPECTROSCOPY

Paul Galland

TABLE OF CONTENTS

I.   Action Spectroscopy 38
     A.   Classical Action Spectroscopy.. 38
     B.   Physical Principles 38
     C.   Measurement of Action Spectra 40
     D.   Reciprocity 43
     E.   Interpretation of an Action Spectrum 44
     F.   Measurement of Action Spectra without Fluence-Response
        Curves.. 44

II.   Kinetic Models for Single-Photoreceptor Systems 45
     A.   Model 1: Photoproduct Formation...................................... 45
     B.   Model 2: Photocatalysis 46

III.   Complex Photoreceptors 46
     A.   Action Spectra of Accessory Pigments 46
     B.   Action Spectra of Photochromes 47
     C.   Photochrome Catalysis 47
     D.   Other Photoreversible Systems 48

IV.   Experimental Approaches Used in Combination with Action
     Spectroscopy... 49
     A.   Quenchers of Excited States.. 49
     B.   Orientation of the Dipole Transition Moment 49
     C.   Behavioral Mutants.. 50
     D.   Use of Photoreceptor Analogs 51

Acknowledgment.. 51

References.. 51

I. ACTION SPECTROSCOPY

Action spectroscopy is the method used to determine in vivo or in vitro the absorption spectrum of the photoreceptor (receptor pigment) that mediates a biological response. The technique consists, in principle, of studying the dependence of the response on wavelength. When the photoreceptor for a particular response is not yet isolated and characterized in vitro, an action spectrum gives the first essential information; i.e., the shape of its absorption spectrum. The action spectrum thus becomes a prerequisite for the future isolation of the photoreceptor. If a specific kinetic model is assumed to hold, action spectroscopy can also give information about the kinetic properties of the receptor system. This is the domain of analytical action spectroscopy, a term coined by Hartmann.[1,2] The reader who intends to study action spectroscopy in greater detail, especially mathematical models, is referred to References 1 through 4.

A. Classical Action Spectroscopy

"Classical" action spectroscopy applies to systems in which a biological response is under the control of a single photoreceptor pigment. An action spectrum is a plot of the reciprocal of the moles of photons causing a defined biological response vs. the wavelength. When the necessary conditions for action spectroscopy as detailed below are met, the action spectrum resembles the absorption spectrum of the photoreceptor. If the optical properties of the system such as screening pigments and scatter are known, the absorption spectrum of the photoreceptor can be measured with great accuracy by action spectroscopy. For a better understanding of action spectroscopy the following paragraph describes a few basic principles about the energetic states of molecules after photon absorption.

B. Physical Principles

According to the Grotthus-Draper law of photochemistry, only photons that are absorbed can cause a photochemical reaction and thereby mediate a biological response. The number of photoreceptor molecules that are transformed is proportional to the number of absorbed photons (Stark-Einstein law of equivalence). The rate constant k with which the photoreceptor is transformed into an excited state is given by:

$$k = N_\lambda \epsilon_\lambda \phi_\lambda \qquad (1)$$

where N_λ is the photon fluence rate (mole$_{(photons)}$ · m^{-2} · sec^{-1}), ϵ_λ the molar extinction coefficient [mole$_{(receptor)}$/m^3]$^{-1}$ m^{-1} and ϕ_λ = mole$_{(receptor)}$/mole$_{(photons)}$ the quantum efficiency. Because the quantum efficiency can be smaller than one, not every absorbed photon leads necessarily to a phototransformation. The product $\epsilon_\lambda \phi_\lambda = \sigma_\lambda$ is called the molar absorption cross-section (m^2 mol^{-1}). It is a measure (with respect to 1 mole of receptor molecules) for the collision area between a photon and the molecule. This collision area (or capture cross-section) reaches only up to 0.5% of the actual cross-section of the molecule. The molar extinction coefficient ϵ_λ can be converted into the collision area q_λ by the following relation:

$$q_\lambda = 0.38 \times 10^{-24} \epsilon_\lambda \ m^2 \qquad (2)$$

The fluence rate f_λ (W·m^{-2}) of a monochromatic light source can be converted into photon-fluence rate N_λ(mol m^{-2} sec^{-1}) by the relation: $f_\lambda \times 8.358 \times \lambda \times 10^{-9} = N_\lambda$. This relation can be easily derived from Einstein's famous formula $E = h\nu$ which states that the energy of an electromagnetic wave is proportional to the product of Planck's constant h and its frequency ν.

The Jablonsky diagram of Figure 1 shows schematically the events that can occur after

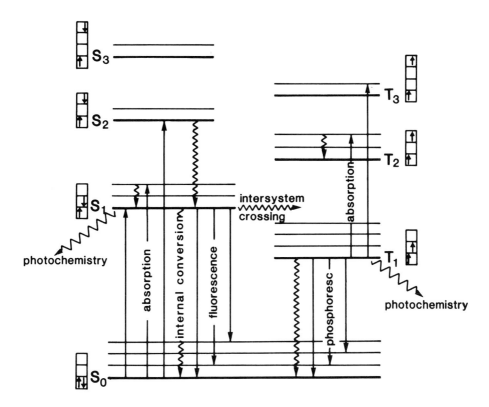

FIGURE 1. Jablonski diagram of an organic molecule. The scheme shows the possible electronic relaxations from the various vibrational levels. The molecule can be excited from the ground state S_0 to the singlet states S_1, S_2, or higher states. From the higher singlet states the molecule can relax to the S_1 state via internal conversion (wavy lines). From the S_1 state the molecule can relax to the ground state S_0 without radiation (internal conversion), or via fluorescence. It can also undergo intersystem crossing to a triplet state, in this case the first triplet state T_1. From the triplet state the molecule can relax either via phosphorescence (phosphoresc) or via internal conversion. The direct relaxation from S_2 (or higher singlet states) to the ground state S_0 is possible but occurs with a probability that is very small compared to the relaxation via the S_1 singlet state. The photochemistry, i.e., the primary reaction of the exited photoreceptor molecule, can occur from the S_1 state or from the T_1 state. The spin of the electrons is indicated by the small arrows in the boxes beside the energy levels.

the absorption of a photon. After photon absorption the molecule goes from the ground state S_0 into one of the excited states. For biological photoreceptor molecules, this usually means that a π-electron goes into one of the excited singlet states S_1, S_2, etc. These transitions are detected in absorption and action spectra. The absorption bands for the triplet transitions T_1 to T_2, etc. are not detectable in conventional spectroscopy and action spectra. An in-vivo role for the higher triplet transitions is, however, theoretically possible under certain conditions and has been proposed once.[29] Usually, however, they can be neglected for action spectroscopy. The S_0 - T_1 transition is quantum mechanically "forbidden"; i.e., the direct excitation of a molecule to the T_1 state is very unlikely because of the requirement of changing the spin and the energy level of an electron simultaneously. This transition is therefore undetectable in conventional absorption spectra. It can, however, be detected by action spectroscopy provided that the photochemistry (primary reaction) occurs from the T_1 state (see below).[18]

From the excited singlet states the molecule can return to the ground state via different routes: (1) through internal conversion; i.e., nonradiative transfer, (2) fluorescence, (3) intersystem crossing from singlet to triplet state followed by phosphorescence or nonradiative transfer, and (4) photochemistry; i.e., by interacting chemically with another molecule. The first molecular change that occurs after the absorption of a photon and that leads to the response is called the *primary reaction*. In principle, the primary reaction could be initiated from any of the excited states. It is, however, most likely that the primary reaction occurs from the lowest energetic states S_1 and T_1, because these have much longer lifetimes than the higher singlet states. S_1 and T_1 therefore have a higher probability of interacting with other molecules eliciting thereby a biological response. If an action spectrum resembles the absorption spectrum of its receptor molecule, it follows that the primary reaction is mediated either by the lowest singlet state S_1 or by the lowest triplet state T_1. An excited molecule can reach S_1 and T_1 from the higher energy states via radiationless deexcitation as shown in Figure 1. On the other hand, if an action spectrum does not resemble the absorption spectrum of the putative photoreceptor, it does not imply that the primary reaction goes via other states than S_1 and T_1. Absence of similarity between the two spectra can have other reasons (discussed below) for complex photoreceptors. In order to do action spectroscopy it is not necessary to know the physical nature of the primary reaction. In fact it is known only for a few well studied cases.

In action spectroscopy the quantum efficiency (also called quantum yield) is defined as the probability that an absorbed photon causes the primary reaction. This probability can often be much smaller than unity. Its magnitude is the product of the quantum efficiency for phototransformation as defined in Equation 1 and of the efficiency with which the primary reaction can compete with the other two processes of deexcitation; i.e., fluorescence (or phosphorescence if the primary reaction occurs from the triplet state T_1) and radiationless deexcitation. The quantum efficiency is considered wavelength independent. This means that the primary reaction always competes with the same two deexcitation processes, independent of which wavelength excited the molecule. The molecular explanation for this assumption is that the highest singlet states first relax to the lower one(s) without also relaxing directly to the ground state (see Figure 1). (Direct relaxation, i.e., spontaneous emission, does actually occur but it is negligible compared to the deexcitation route via S_1). If, for example, the S_2 state in Figure 1 could also relax directly to the S_0 state, the quantum efficiency would be wavelength dependent. This is so because the primary reaction would then compete with different deexcitation routes (those from S_2 and from S_1) depending on the singlet state to which the molecule was excited.

Action spectroscopy is based on the fundamental assumption that the system must give identical responses to different wavelengths when the rates of the primary reactions are identical. The rate of the primary reaction must be proportional to the rate constant of phototransformation in Equation 1. Therefore, if two responses to wavelengths λ_1 and λ_2 are identical, the rate constants of phototransformation must also be identical: $k_{\lambda 1} = k_{\lambda 2}$, or $N_{\lambda 1} \cdot \sigma_{\lambda 1} = N_{\lambda 2} \cdot \sigma_{\lambda 2}$. This leads to the relation:

$$\sigma_{\lambda 1}/\sigma_{\lambda 2} = 1/N_{\lambda 1} : 1/N_{\lambda 2} \tag{3}$$

This basic equation of action spectroscopy states that the cross-sections of phototransformation are proportional to the reciprocals of the photon fluence rates.

C. Measurement of Action Spectra

In practice, an action spectrum is measured with the following two-step procedure. First one measures the response as a function of N_λ, the photon-fluence rate (mol m^{-2} sec^{-1}). When light pulses rather than continuous irradiation are used to elicit the response, N_λ should be replaced by $F_\lambda = N_\lambda \cdot \Delta t$, the photon fluence (mol m^{-2}) where Δt is the duration of the light pulse with photon fluence rate N_λ. The equations that are used to construct an action spectrum apply equally well to F_λ and N_λ. For any given wavelength one measures these

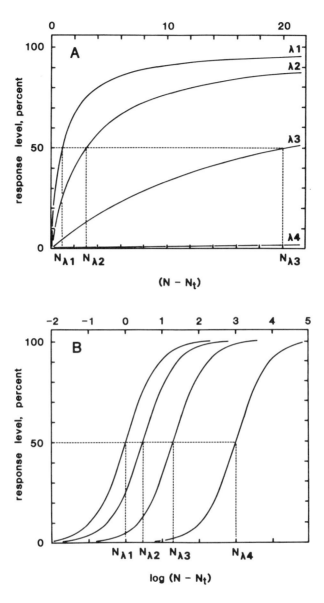

FIGURE 2. Photon fluence rate-response curves for wavelengths of different effectiveness. The level of a hypothetical biological response is plotted vs. the photon fluence rate N ($\text{mole}_{\text{photons}}$ m^{-2} $\sec^{-1}$). (A) Linear plot; (B) logarithmic plot. Because the threshold N_t (fluence rate or fluence) for a response is specific for the organism, the conditions used and the type of response, N_t was subtracted from the fluence rate N. In this way the curves in the linear plot (A) pass through zero. The amounts by which the values $(N - N_t)$ that elicit the criterion response are shifted along the abscissa correspond to the decrease in the relative cross sections of the photoreceptor pigment in Figure 3A and B.

photon-fluence rate-response curves as shown in Figure 2. In order to compare the efficiencies of the various wavelengths, all other parameters of the system, such as the concentration of the photoreceptor, temperature, time schedule of irradiation, and physiological conditions of the organism have to stay constant. A criterion response (for instance a 50% response level as in Figure 2) is chosen and the photon fluence rate $N_{\lambda i}$ at λ_i that is required to elicit

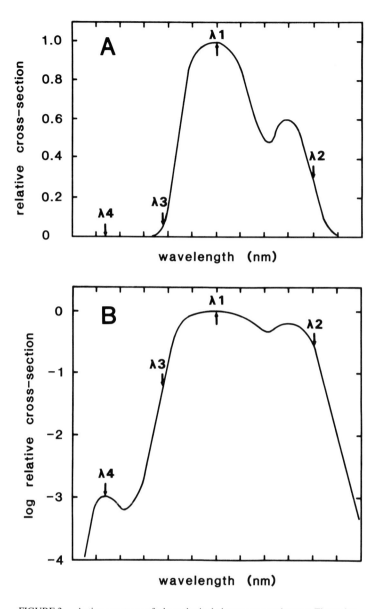

FIGURE 3.   Action spectrum of a hypothetical photoreceptor pigment. The action spectrum is a plot of 1/N (obtained from photon fluence response curves as shown in Figure 2) vs. wavelength. In this action spectrum the ordinate shows 1/N standardized with respect to $1/N_{ref}$ of the reference wavelength. The wavelength with the smallest photon requirement is chosen as the reference wavelength. (A) Linear scale; (B) same action spectrum in logarithmic scale.

the response is determined. A plot of $1/N_\lambda$ vs. wavelength is the action spectrum. Alternatively, one may normalize all values of $1/N_{\lambda i}$ to the value of $1/N_{\lambda ref}$ for a reference wavelength, usually the wavelength where photon requirement is the lowest. This has been done in Figure 3A and B with the action spectrum of a hypothetical photoreceptor. Because $1/N_\lambda$ is proportional to $\sigma_\lambda = \epsilon_\lambda \phi_\lambda$ the action spectrum represents the relative spectrum of the cross-section of the photoreceptor. To obtain the absolute spectrum of the cross-section, one would have to know the quantum efficiency of phototransformation (Equation 1), as well as the contribution of screening that reduces the efficiency of the irradiation. Once the photoreceptor and its molar extinction coefficient ϵ_λ are known the quantum efficiency can be obtained from fluence response curves.

In the literature a great number of different names for the ordinate of action spectra can be found. Common designations include relative cross-section, relative photon effectiveness, photon responsivity, relative quantum effectiveness, relative quantum efficiency, relative effectiveness, and relative quantum responsivity. It has become customary to designate the ordinate of an action spectrum ''relative quantum efficiency'' (or ''relative quantum effectiveness'') instead of relative cross-section. This is somewhat misleading because the quantum efficiency shown on the ordinate of action spectra usually refers to response per number of incident quanta. This is not identical with the quantum efficiency ϕ_λ of phototransformation in the formula $\sigma_\lambda = \epsilon_\lambda \phi_\lambda$, where it is by definition wavelength independent (see above).

In Figures 2 and 3 the two-step procedure is shown in linear as well as logarithmic plots. Both plots contain of course the same information. In practice, however, it is of advantage to use the logarithmic plot for the photon fluence rate-response curves (or photon fluence-response curves). Besides the obvious advantage of enabling one to plot a wide range of photon fluence rates, this method is particularly useful because for any two wavelengths that bring about the same response the curves must be shifted parallel to each other by the interval $\ln \sigma_{\lambda 2}/\sigma_{\lambda 1}$, as $\ln N_{\lambda 1} - \ln N_{\lambda 2} = \ln \sigma_{\lambda 2}/\sigma_{\lambda 1}$.

Finding parallel photon fluence-response curves thus provides a test for the underlying assumptions of classical action spectroscopy. If the photon fluence rate-response curves (or photon fluence-response curves) are not parallel, one deals most likely with a more complex photoreceptor system as will be discussed below.

Using a logarithmic ordinate for the relative cross-section can be of advantage, as shown for the hypothetical action spectrum in Figure 3B. A very small peak (at λ_4) of a relative cross-section of 10^{-3} does not appear in a linear plot (Figure 2B) but becomes visible in the logarithmic plot. Such small peaks can sometimes be caused by artifacts unless great care is taken to insure the purity of the light source.

D. Reciprocity

Equation 3 states that $1/N_{\lambda i}$ is proportional to the cross-section of the photoreceptor molecule at λ_i. This equation is the only condition for classical action spectroscopy. It does not make any statement about the time schedule of irradiation as long as the irradiation times are constant for the different wavelengths. Often the response depends on the product of $N \times \Delta t$; i.e., it is independent of the single values of N and Δt as long as the product of the two parameters is constant: $N \times \Delta t = $ const. The system is then said to obey the Bunsen-Roscoe law of reciprocity.[51] If an action spectrum is measured on the basis of fluence-response curves, i.e., with pulses, one has to check first in which range of Δt and N reciprocity holds. In this range only Δt and N may be varied. Only then one can legitimately compare fluence-response curves for different wavelengths that were measured with different pulse durations Δt and photon fluence-rates. In practice this problem often occurs, because of the strong wavelength dependence of the energy output of light sources. Incandescent lamps, for example, contain relatively little ultraviolet light. At those wavelengths where the energy output of the lamps becomes limiting, higher fluences are often obtained by applying much longer pulses than the ones used at wavelengths where the lamp has a high energy output. If this is done, it has first to be assured that the limits of reciprocity are not violated. This is the reason why validity of reciprocity is often said to be a prerequisite for action spectroscopy. Classical action spectroscopy, though, is certainly not meaningless outside the range where reciprocity holds. Equation 3 does not imply that. If action spectra are measured for pulse durations or photon fluence rates outside the range of reciprocity, it is enough to be sure that these parameters (Δt and N) are kept constant for all wavelengths. Only if these two parameters are changed for different wavelengths does one have to work within the range of reciprocity.

In a single-photoreceptor system, reciprocity must be valid at any wavelength for constant threshold fluences (where only a small fraction of the photoreceptor population is exited) and pulse durations that are short in comparison to the actual response. Reciprocity failure under these conditions indicates a complex photoreceptor and shows that the conditions for classical action spectroscopy are not met. An example is the phototropic response of *Phycomyces* where reciprocity breaks down near the threshold fluence when the light pulses become shorter than 67 msec.[6]

E. Interpretation of an Action Spectrum

Finding that an action spectrum is identical to the absorption spectrum of a known molecule makes it likely, but does not prove, that this molecule functions as the photoreceptor for the response under study. Identity of action spectra and absorption spectra can be expected if the screening of other pigments and substances in the cell is wavelength independent. In practice, however, this is never the case, because all cells contain some screening materials. In addition the photoreceptor itself can contribute to the overall absorption of the cell by self-screening. For these reasons it is of advantage to measure the absorption spectrum of the cell or tissue in order to correct the action spectrum for these screening effects. However, even if absorption spectra of the screening materials are known, it is not always possible to estimate the true optical path-length in the vicinity of the photoreceptor. Most experimenters try to minimize these artifacts by working with thin samples. For experimentation with thick and turbid samples it is often necessary to apply mathematical models with which one can estimate scatter, screening, etc.[3,7,8]

The impact of passive screening pigments on the shape of an action spectrum can sometimes be controlled by the use of mutants that are lacking those pigments (see below) or else with compounds that inhibit their synthesis. Herbicides that inhibit chlorophyll synthesis in higher plants are frequently used for phytochrome studies because of the overlapping absorption spectra of these two receptor molecules.

The action spectrum does not necessarily resemble the absorption spectrum of the chromophore of the photoreceptor pigment because the absorption spectra of the chromophore and the whole photoreceptor pigment (protein moiety plus chromophore) can greatly differ. Well known examples are the rhodopsin pigments. Bovine rhodopsin has a major absorption peak at 500 nm while its chromophore, 11-cis retinal, absorbs at 370 nm.

Even when all the conditions for classical action spectroscopy are met, the action spectrum does not necessarily describe a single photoreceptor pigment. In a system with two photoreceptors that contribute additively to a single primary reaction, the conditions for classical action spectroscopy including validity of reciprocity are still valid. Therefore, *in sensu strictu*, an action spectrum gives the relative cross-section of an *effective* receptor pigment. This "pigment" may represent the contribution of more than one molecular species.

F. Measurement of Action Spectra without Fluence-Response Curves

In some special cases action spectra can be obtained without the tedious measurement of photon fluence-rate (or photon fluence) response curves. This has been done, for instance, with action spectra of photogeotropism of *Phycomyces* sporangiophores.[9] In this technique the phototropic organ is in balance between two light sources: the reference wavelength of constant photon fluence rate and the test wavelength of variable photon fluence rate. If only one photoreceptor mediated phototropism, one would obtain similar action spectra with the two different techniques. However, in *Phycomyces* the balance method can give action spectra that are substantially different from action spectra that are based on photon-fluence rate-response curves (monochromatic unilateral light). This indicates most likely a complex photoreceptor system.[9,10]

Published action spectra are often not based on the measurement of photon fluence rate-response curves (or photon fluence-response curves) or on the balance method as described above. Instead, the response of the system to a constant number of photons is measured. This very popular shortcut to the measurement of action spectra in the proper but laborious way is based on the assumption that the fluence-response curves are linear functions. In reality, fluence-response curves are either exponential or hyperbolic (or still more complex). As a result this method results in an artificial peak broadening and should be avoided. Additionally, this method ignores the fact that the slope of fluence-response curves (plotted on a logarithmic abscissa as in Figure 2B) is often wavelength dependent, in which case the artifacts are aggravated.

II. KINETIC MODELS FOR SINGLE-PHOTORECEPTOR SYSTEMS

Besides information about the spectral properties of a photoreceptor, fluence-response curves can also give information about the kinetic properties of a receptor system. This becomes possible in the framework of specific kinetic models and is part of analytical action spectroscopy. In the case of a single photoreceptor two systems have been distinguished on the basis of kinetic properties. A detailed description of these mathematical models by Hartmann can be found in References 1 and 2.

A. Model 1: Photoproduct Formation

In this model it is assumed that upon absorption of light the photoreceptor P goes into an excited state A which is subsequently converted into a product R by which the biological reaction is elicited:

$$\overset{h\nu}{\rightsquigarrow} P \underset{k_A}{\overset{k_{A\lambda}}{\rightleftarrows}} A \overset{k}{\rightarrow} B \rightarrow \rightarrow \ldots R$$

It can be shown that the relative effect $y_{rel} = R/R_{max}$ becomes

$$y_{rel} = 1 - \exp(-\sigma_\lambda \cdot N_\lambda \cdot t) \qquad (4)$$

Thus the relative effect y_{rel} depends exponentially on the product N (photon fluence rate) and the exposure time t. The photon fluence rate N and the exposure time t can be exchanged which fulfills the requirement for reciprocity. This model predicts exponential photon fluence-response curves for monochromatic as well as polychromatic irradiation. Because the model is based on the assumption of a single photoreceptor and a single primary reaction the effect of the polychromatic irradiation can be predicted from the responses to monochromatic irradiation. One should therefore check experimentally with dichromatic irradiation whether or not the following condition is met:

$$y_{rel} = 1 - \exp - (\sigma_{\lambda 1} \cdot N_{\lambda 1} \cdot t + \sigma_{\lambda 2} \cdot N_{\lambda 2} \cdot t) \qquad (5)$$

If this condition is not fulfilled the model can be ruled out. An example for photoproduct formation is the light-dependent synthesis of chlorophyll in higher plants.[11]

Another example is the photoinactivation by ultraviolet light of viruses, bacteriophages, and enzymes. In this case it is customary to analyze the particles remaining after irradiation (survivors) rather than the inactivated fraction. One analyzes the plots $1 - y_{rel} = P_t/P_0$ where P_t is the fraction of P not hit by photons at time t and P_0 the fraction of P at time t = 0. Exponential photon fluence response curves and reciprocity are expected in this case for single-hit processes; i.e., when the particle is inactivated by a single absorbed photon.

B. Model 2: Photocatalysis

In this model the active state A of the photoreceptor P serves as a catalyst and is not transformed into other products. A is phototransformed into A and A reverts in the dark to the inactive form P with the rate constant k_A:

$$\overset{h\nu}{\rightsquigarrow} P \underset{k_A}{\overset{k_{A\lambda}}{\rightleftharpoons}} A \quad k \quad \cdots y$$

For this model, it was shown[1,2] that the response y becomes

$$y = k \cdot P \cdot t[1 + k_A/(\sigma_{A\lambda} \cdot N_\lambda)]^{-1} \tag{6}$$

Thus y depends linearly on the exposure time t but hyperbolically on the photon-fluence rate N_λ. The Bunsen-Roscoe law of reciprocity is therefore not valid. If the exposure times are kept constant for all wavelengths, the relative effect y_{rel} becomes:

$$y_{rel} = [1 + k_A/(\sigma_A \cdot N)]^{-1} \tag{7}$$

Thus one obtains hyperbolical photon fluence response curves. Even in this system the reaction to polychromatic irradiation is in principle predictable by the reactions to monochromatic light. One should therefore test with dichromatic irradiation whether or not the following condition is met:

$$[y_{rel}/(1 - y_{rel})]_{\lambda 1 + \lambda 2} = [y_{rel}/(1 - y_{rel})]_{\lambda 1} + [y_{rel}/(1 - y_{rel})]_{\lambda 2}$$

$$= \sigma_{A\lambda 1} \cdot N_{\lambda 1}/k_A + \sigma_{A\lambda 2} \cdot N_{\lambda 2}/k_A \tag{8}$$

Examples for photocatalyses are the photoactivation of the carbon monoxide inhibited cytochrome oxidase[12] and the photoactivation of nitrate reductase.[13] Nitrate reductase transfers electrons from photoreduced riboflavin to nitrate, reducing it to nitrite.

III. COMPLEX PHOTORECEPTORS

As shown above, classical action spectroscopy can be applied only to systems with a single photoreceptor and a single primary response. It is furthermore essential that the photoreceptor concentration and other physiological parameters of the system are constant throughout the time of irradiation. Numerous complex photoreceptor systems are known for which these conditions do not apply. Examples are the photoreceptor of photosynthesis and their accessory pigments in higher plants, and photochromic photoreceptors such as invertebrate rhodopsin and phytochrome. Even though the basic conditions for classical action spectroscopy are not fulfilled in these cases, action spectroscopy is not useless. Some of the complications that at first glance appear to hinder the analysis actually lend themselves to the identification of the receptors. The exact absorption spectrum of phytochrome, for example, cannot be determined by classical action spectroscopy alone, but the photoreversibility of phytochrome was an excellent tool for its identification and in vitro isolation.

A. Action Spectra of Accessory Pigments

The photosynthetic effectiveness of far-red light (>675 nm) is smaller than expected from the absorption spectrum of chlorophyll. This is called the "red drop". When far-red light is given simultaneously with shorter-wavelength irradiation, the photosynthetic effectiveness is greatly enhanced. With double irradiations of this type one can show that the validity of

Equation 8 (see above) does not hold. The explanation for this effect (second Emerson effect) is that the essential photosynthetic accessory pigments absorb at shorter wavelength. Action spectra for the enhancement of the far-red light efficiency showed the presence of accessory pigments.[14]

Another curiosity that allowed spectroscopy of the accessory pigments in photosynthesis is the occurence of "chromatic transients".[14] If the oxygen evolution in algae is adjusted to an equal rate with either red or green light, the sudden switch from red to green or vice versa results in a transient fluctuation of the oxygen evolution rate even though both single wavelengths are subjectively equivalent.[14] It appears as if under these conditions the system had color discrimination, a property which requires more than on photoreceptor. Action spectra for these transients are very similar in shape to the ones for the enhancement. Experiments of this type led to the concept of accessory pigments and two photoacts in photosynthesis.[14]

It should be clear at this point that such complications could not occur in a single-photoreceptor system.

B. Action Spectra of Photochromes

A photochrome is a molecule whose absorption spectrum is modified after absorbing a photon. In the simplest case the molecule goes from a state A into a state B from which it can revert to A in a dark reaction:

$$A \underset{}{\overset{h\nu}{\rightleftarrows}} B$$

The reversion from B to A might also depend on light, i.e., the system can be photoreversible:

$$A \underset{h\nu_2}{\overset{h\nu_1}{\rightleftarrows}} B$$

If the absorption spectra of A and B overlap, classical action spectroscopy does not apply because light elicits two primary reactions with opposing effects.

The most thoroughly studied example of a photochrome is the plant pigment phytochrome.[15] Phytochrome exists in darkness in a red-light-absorbing form P_{660} (P_r) which undergoes after absorption of red light a transition into a far-red-light-absorbing form, P_{730} (P_{fr}): $P_r \underset{730 \text{ nm}}{\overset{660 \text{ nm}}{\rightleftarrows}} P_{fr}$, with P_{fr} as the biological effector molecule. Alternating short-term irradiation with red light (660 nm) and far-red light (730 nm) leads to response induction and response reversal, respectively, depending which wavelength is given last. Action spectra that are measured under these inductive conditions have peaks around 660 nm for the induction response and around 730 nm for the response reversal.[15] Under continuous irradiation, however, the shape of the phytochrome action spectra is very different, among other reasons because of the instability of the P_{fr} molecule. The interpretation of those action spectra on the exclusive basis of phytochrome action requires special assumptions about the kinetic properties and stability of the phytochrome receptor and can most likely not be generalized for all photochromes.[3] A general model for the action of phytochrome under inductive and steady-state conditions was proposed by Schäfer.[16]

C. Photochrome Catalysis

A general kinetic model for photochromes was proposed by Hartmann, and can be found in detail in References 1 and 2. According to this model the photoreceptor P undergoes a phototransformation into the effector form A that acts as a catalyst. A can photorevert and also dark revert into the biologically inactive receptor from P:

$$\overset{h\nu}{\rightsquigarrow} P \underset{k_{pa} + k_A}{\overset{k_{A\lambda}}{\rightleftharpoons}} A \quad k \quad \cdots \quad y$$

In darkness the entire photochrome exists in the P form. For short light exposures, i.e., $t \rightarrow 0$, dA becomes $dA = P \cdot \sigma_{A\lambda} \cdot E_\lambda \cdot t$. Therefore, for small t, i.e., for the initial phase of the photon fluence response curves (relative response $y \rightarrow O$), the relative photon effectiveness gives the relative absorption spectrum of the P form (the relative apparent conversion spectrum for the effector formation $P \rightarrow A$). The photon flux dependency in this initial phase is exponential and reciprocity is valid.

For longer exposure times, the photoreversion from A to P comes into play. If the exposure times are chosen such that they are small relative to the inactivation time $1/k_A$, k_A can be neglected. The saturation plateaus of the photon fluence response curves become wavelength dependent in this condition and it is proportional to the photo-steady-state fraction of the effector $A/(A + P) \propto \sigma_A/(\sigma_A + \sigma_P)$. The relative photon effectiveness p_{rel} becomes:

$$p_{rel} = N_{\lambda 2} \cdot t/N_{\lambda 1} \cdot t = (\sigma_{A\lambda 1} + \sigma_{P\lambda 1})/(\sigma_{A\lambda 2} + \sigma_{P\lambda 2}) \tag{9}$$

Thus, the action spectrum in this condition gives the relative sum of the apparent conversion spectra (absorption spectra) of both the P and the A form of the photochrome.

For long-term or continuous exposure it can be shown that the relative effect y_{rel} in photo-steady state becomes:

$$y_{rel} = y/y_{sat} = (k_{A\lambda} + k_{P\lambda})/(k_{A\lambda} + k_{P\lambda} + k_A) \tag{10}$$

The saturation level y_{sat} depends on the photon flux ratio of $N_{\lambda 1}/N_{\lambda 2}$. For dichromatic irradiation the following equation holds:

$$[y_{rel}/(1 - y_{rel})]_{\lambda 1 + \lambda 2} = [y_{rel}/(1 - y_{rel})]_{\lambda 1} + [y_{rel}/(1 - y_{rel})]_{\lambda 2}$$

$$= 1/k_A[(\sigma_{A\lambda 1} + \sigma_{P\lambda 1}) N_{\lambda 1} + (\sigma_{A\lambda 2} + \sigma_{P\lambda 2}) N_{\lambda 2}] \tag{11}$$

This kinetic photochrome model of Hartmann does not rest on the assumption that the response should be revertible by alternating short-term irradiation. A response reversal such as found in phytochrome depends on the presence of a substantial delay between the formation of the effector A and the formation of y. Thus absence of photoreversibility with alternating short-term irradiation does not exclude photochromicity. In such a case the photochrome can still be identified in the framework of this kinetic model. This is best done by measuring the relative photon effectiveness once for very small irradiation times (relative response $y \rightarrow 0$), which gives the relative conversion spectrum for the P form of the photochrome ($P \rightarrow A$) and once for longer times where the relative photon effectiveness follows the sum of the conversion spectra $P \rightarrow A$. The saturation level of the photon fluence response curves must be wavelength dependent and depend on the photo-steady-state fraction of the effector $A/(A + P)$.

D. Other Photoreversible Systems

Photoreversibility of a response by alternating irradiation with different wavelength does not necessarily indicate the presence of a photochrome. Photoreversibility can in principle also be explained on the basis of two (or more) independent and antagonistically interacting photoreceptors. Photophysiology and action spectroscopy alone cannot distinguish between these alternatives and one depends therefore also on biochemical data. An example is the photoreceptor system "mycochrome" that reversibly controls conidial development in *Alternaria tomato*. This mycochrome system contains two individual receptors, P_{NUV} and P_B, that can be separated in vitro.[17] Irradiation with near ultraviolet light induces conidiation while a subsequent irradiation with blue light inhibits conidiogenesis.[17] A general theory for those reversible systems has not yet been developed. But with dichromatic irradiations and help of Equations 4 or 7 one should be able to rule out a single-photoreceptor mechanism with a single primary response.

IV. EXPERIMENTAL APPROACHES USED IN COMBINATION WITH ACTION SPECTROSCOPY

Even if action spectroscopy is done very carefully according to the principles of classical and analytical action spectroscopy, it is often necessary to have further information to distinguish between alternative hypotheses. During the last two decades a number of experimental approaches have been frequently adopted that can greatly extend the information obtained from action spectroscopy. In combination with action spectra they are powerful tools to analyze biological photoreceptors.

A. Quenchers of Excited States

The question whether the primary reaction occurs from the lowest triplet state T_1 or from the lowest singlet state S_1 can rarely be decided by action spectroscopy. The direct transition from S_0 to T_1 is quantum mechanically "forbidden" because of the low probability to simultaneously change the spin of the electron and its energy level. Simple absorption spectra of molecules do not have therefore a triplet peak. The peak position of the lowest triplet state has to be determined by phosphorescence. Because of the extremely high fluence rates that are required to directly excite a measurable fraction of molecules to the triplet state, it is usually impractical to detect the expected contribution of the action spectrum. An attempt to detect direct excitation to the lowest triplet state of the flavin photoreceptor was made for the light-growth response of the sporangiophore of the fungus *Phycomyces*.[18]

It is, however, possible to use compounds that specifically quench the singlet or the triplet states of the photoreceptor molecule. The in vivo effect of such compounds can in principle tell whether a primary reaction occurs from the singlet or the triplet state. An example is the heavy atom gas xenon that quenches in vitro the triplet excited state of flavin. However, the phototropic response of corn seedlings, thought to be mediated by a flavin photoreceptor, is unaffected by xenon.[19] On the other hand, azide, an efficient singlet quencher of flavins, is able to inhibit phototropism in corn.[19] These results make it likely that a flavin singlet rather than a triplet state is involved in the early photoprocesses of this response.

B. Orientation of the Dipole Transition Moment

The electronic transitions from the ground state S_0 to the higher states depend not only on the frequency of the photons but also on the orientation of the electrical vector with respect to the molecule. Figure 4 shows the example of a flavin molecule and the orientations of the transition dipole moments for the different wavelengths. The orientation of the transition dipole moment is, like the absorption spectrum, specific for a given molecule and can therefore serve as a means to identify the photoreceptor. The only prerequisite to apply this method is that the photoreceptor molecules be oriented.

Many photoreceptors are not randomly oriented in the cell but have a preferred orientation with respect to the rest of the cell. This so-called dichroism of a photoreceptor is usually

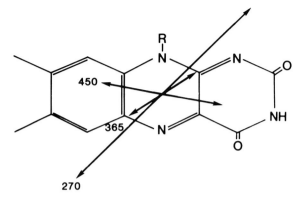

FIGURE 4. Dipole transition moments at 450, 365, and 270 nm for riboflavin taken from fluorescence polarization measurements.[20,21] The length of the arrows indicates the relative intensity of the transitions. R = ribityl moietyCH_2–(H–C––OH)$_3$–CH_2OH. (Reprinted in part from *Photochemistry and Photobiology*, Vol. 7, Kurtin, W. E. and Song, P.-S., Photochemistry of the model phototropic system involving flavins and indoles. I. Fluorescence polarization and M.O. calculations of the direction of the electron transition moments in flavins, copyright 1968, Pergamon Journals, Ltd.)

believed to result from anchoring of the molecules to the matrix of the plasma membrane. Experimentally, photoreceptor dichroism is detected by using linearly polarized light. Dichroism is made manifest by the fact that the threshold for a photoresponse depends on the orientation of the polarized light. For optimal effectiveness, the electrical vector of the polarized light beam has to coincide with the transition dipole moment of the photoreceptor molecule. If action dichroism in vivo has the same wavelength dependence found for the transition dipole moment of the photoreceptor candidate in vitro, the two substances are probably the same. In this way, an independent means of identification of the photoreceptor is obtained.

An example for this approach is the action dichroism found for the light-growth response of the sporangiophore of *Phycomyces*, thought to be mediated by flavin photoreceptors.[22] The relative angles of electrical vectors of the polarized light beams that gave optimal response correspond to the orientation of the transition dipole moments found in vitro for flavins (Figure 4), confirming the flavin nature of the photoreceptor.

C. Behavioral Mutants

The use of biochemcially defined mutants has helped in some cases to identify or to exclude a photoreceptor candidate. In the lower fungus *Phycomyces* it was found that beta-carotene-lacking strains retained full phototropic sensitivity.[23] This finding excluded beta-carotene or any of its metabolic derivatives as photoreceptor candidates. Mutants that lack the ubiquitous pigment beta-carotene are also particularly useful for estimating the contribution of this pigment to screening. Light absorption by passive screening pigments like beta-carotene can reduce the slope of fluence-response curves. Mutants of *Phycomyces* that lack beta-carotene have a slightly steeper slope in the phototropic fluence response curves than the wild type; i.e., require less light for the same response level.[23]

Another example where a metabolic mutant led to the positive (though indirect) identification of the photoreceptor is *Chlamydomonas*. In this unicellular alga, beta-carotene-lacking mutants are phototactically defective.[24] The reason is that *Chlamydomonas* uses retinal, a derivative of beta-carotene, as the chromophore of the photoreceptor. Addition of retinal to those mutants restores their photosensitivity.[24] These results confirm and extend the conclusions drawn from the phototactic action spectra of *Chlamydomonas* that indicate a rhodopsin photoreceptor.[24]

The presence of multiple photoreceptors in *Halobacterium* indicated by a complex photophysiology, was also supported genetically. It was possible to genetically distinguish three photoreceptors that otherwise would have been difficult to demonstrate by exclusively classical physiological and spectroscopic means.[25,26]

Another example are behavioral mutants of *Arabidopsis thaliana* that are defective in the light inhibition of hypocotyl lengthening.[27] From this study it was concluded on purely genetic grounds that the photoinhibition must be controlled by more than just one photoreceptor. Traditionally the inhibition of hypocotyl lengthening is ascribed to the action of phytochrome.

D. Use of Photoreceptor Analogs

Another approach that is used more and more frequently takes advantage of photoreceptor analogs that have a different absorption spectrum than the would-be photoreceptor. The rationale is that an analog of the photoreceptor candidate should be incorporated *in situ* into the chromophore site, thus modifying the action spectrum in accordance with its own modified absorption spectrum. This approach was employed successfully in the case of the phototaxis of *Chlamydomonas*.[24] Retinal analogs with a spectral shift in the absorption spectrum caused a corresponding shift in the action spectrum of phototaxis, showing that retinal must be the chromophore of a rhodopsin receptor for phototaxis.[24] Another example of this approach is *Phycomyces* phototropism: roseoflavin, a riboflavin analog with a bathochromic (i.e., toward longer wavelength) shift of the absorption spectrum, can cause a similar shift in the phototropic action spectrum, showing that a flavin is likely to function as photoreceptor in phototropism.[28] On the other hand, if the analog does not induce the anticipated shift in the action spectrum, the negative result does not rule out the photoreceptor candidate.

ACKNOWLEDGMENT

I am indebted to Drs. Lipson and Horwitz for critically reading this manuscript and for numerous suggestions.

REFERENCES

1. **Hartmann, K. and Cohnen-Unser, I.,** Analytical action spectroscopy with living systems: photochemical aspects and attenuance, *Ber. Dtsch. Bot. Ges.,* 85, 481, 1972.
2. **Hartmann, K.,** Aktionsspektroskopie, in *Biophysik,* Hoppe, W., Lohmann, W., Markl, H., and Ziegler, H., Eds., Springer-Verlag, Berlin, 1977, 197; English edition, Springer-Verlag, Berlin, 1983, 115.
3. **Schäfer, E., Fukshansky, L., and Shropshire, W., Jr.,** Action spectroscopy of photoreversible systems, in *Encyclopedia of Plant Physiology,* New Series, Vol. 16A, Shropshire, W., Jr. and Mohr, H., Eds., Springer-Verlag, Berlin, 1983, 39.
4. **Fukshansky, L. and Schäfer, E.,** Models in photomorphogenesis, in *Photomorphogenesis,* Shropshire, W., Jr. and Mohr, H., Eds., *Encyclopedia of Plant Physiology,* New Series, Vol. 16A, Shropshire, W., Jr. and Mohr, H., Eds., Springer-Verlag, Berlin, 1983, chap. 5.
5. **Bunsen, R. W. and Roscoe, H.,** Photochemische Untersuchungen. VI. Meteorologische Lichtmessungen, *Ann. Phys. Chem.,* 117, 529, 1862.
6. **Galland, P., Palit, A., and Lipson, E. D.,** *Phycomyces:* phototropism and light-growth response to pulse stimuli, *Planta,* 165, 538, 1985.
7. **Butler, W. R.,** Absorption of light by turbid materials, *J. Opt. Soc. Am.,* 52, 292, 1962.
8. **Fukshansky, L. and Kazarinova, N.,** Extension of the Kubelka-Munk theory of light propagation in intensely scattering materials to fluorescent media, *J. Opt. Soc. Am.,* 70, 1101, 1980.

9. **Galland, P. and Lipson, E. D.,** Action spectra for phototropic balance in *Phycomyces:* dependence on reference wavelength and intensity range, *Photochem. Photobiol.,* 41, 323, 1985a.

10. **Galland, P. and Lipson, E. D.,** Modified action spectra of photogeotropic equilibrium in *Phycomyces* mutants with defects in genes *madA, madB, madC,* and *madH, Photochem. Photobiol.,* 41, 331, 1985b.

11. **Koski, V. M., French, C. S., and Smith, J. H. C.,** The action spectrum of protochlorophyll to chlorophyll a in normal and albino corn seedlings, *Arch. Biochem. Biophys.,* 31, 1, 1951.

12. **Warburg, O.,** Das sauerstoffübertragende Ferment der Atmung, *Angew. Chem.,* 45, 1, 1932.

13. **Zumft, W. G., Castillo, F., and Hartmann, K. M.,** Flavin-mediated photoreduction of nitrate by nitrate reductase of higher plants and microorganisms, in *The Blue Light Syndrome,* Senger, H., Ed., Springer-Verlag, Berlin, 1980, 422.

14. **Blinks, L. R.,** Accessory pigments and photosynthesis, in *Photophysiology,* Vol. 1, Giese, A. C., Ed., Academic Press, New York, 1964, chap. 7.

15. **Shropshire, W., Jr.,** Photomorphogenesis, in *The Science of Photobiology,* Smith, K. C., Ed., Plenum Press, New York, 1977, chap. 11.

16. **Schäfer, E.,** A new approach to explain the "high irradiance responses" of photomorphogenesis on the basis of phytochrome, *J. Math. Biol.,* 2, 41, 1975.

17. **Kumagai, T.,** Blue and near ultraviolet reversible photoreaction in conidial development of certain fungi, in *The Blue Light Syndrome,* Senger, H., Ed., Springer-Verlag, Berlin, 1980, 251.

18. **Delbrück, M., Katzir, A., and Presti, D.,** Responses of *Phycomyces* indicating optical exitation of the lowest triplet state of riboflavin, *Proc. Natl. Acad. Sci. U.S.A.,* 73, 1969, 1976.

19. **Vierstra, R. D., Poff, K. L., Walker, E. B., and Song, P.-S.,** Effect of xenon on the excited states of phototropic receptor flavin in corn seedlings, *Plant. Physiol.,* 67, 996, 1981.

20. **Kurtin, W. E. and Song, P.-S.,** Photochemistry of the model phototropic system involving flavins and indoles. I. Fluorescence polarization and MO calculations of the direction of the electronic transition moments in flavins, *Photochem. Photobiol.,* 7, 263, 1968.

21. **Siódmiak, J. and Frackowiak, D.,** Polarization of fluorescence of riboflavin in anisotropic medium, *Photochem. Photobiol.,* 16, 173, 1972.

22. **Jesaitis, A. J.,** Linear dichroism and orientation of the *Phycomyces* photopigment, *J. Gen. Physiol.,* 63, 1, 1974.

23. **Presti, D., Hsu, W.-J., and Delbrück, M.,** Phototropism in *Phycomyces* mutants lacking β-carotene, *Photochem. Photobiol.,* 26, 403, 1977.

24. **Foster, K. W., Saranak, J., Patel, N., Zarilli, G., Okabe, M., Kline, T., and Nakanishi, K.,** A rhodopsin is the functional photoreceptor for phototaxis in the unicellular eukaryote *Chlamydomonas, Nature,* 311, 756, 1984.

25. **Spudich, J. L.,** Genetic demonstration of a sensory rhodopsin in bacteria, in *Information and Energy Transduction in Biological Membranes,* Colombetti, G. and Lenci, F., Eds., Alan R. Liss, New York, 221, 1984.

26. **Sundberg, S. A. and Spudich, J. L.,** Genetic analysis of halobacterial phototaxis, *Biophys. J.,* 47, 201a, 1985.

27. **Koorneef, M., Rolff, E., and Spruit, C. J. P.,** Genetic control of light-inhibited hypocotyl elongation in *Arabidopsis thaliana* (L.) Heynh., *Z. Pflanzenphysiol.,* 100, 147, 1980.

28. **Otto, M. K., Jayaram, M., Hamilton, M., and Delbrück, M.,** Replacement of riboflavin by an analogue in the blue-light photoreceptor of *Phycomyces, Proc. Natl. Acad. Sci. U.S.A.,* 78, 266, 1981.

29. **Schmidt, W., Hart, J., Filner, P., and Poff, K. L.,** Specific inhibition of phototropism in corn seedlings, *Plant Physiol.,* 60, 736, 1977.

Chapter 4

FIRST MEASURABLE EFFECTS FOLLOWING PHOTOINDUCTION OF MORPHOGENESIS

Benjamin A. Horwitz and Jonathan Gressel

TABLE OF CONTENTS

I.   Introduction 54
        A.   Early Effects and the Transduction Chain 54
        B.   Biochemical Development 54
        C.   Transcription/Translation and Morphogenesis without Blue
              Light 54
        D.   Rapid Effects 56

II.   Transcription and Translation 57
        A.   Cellular (Internal) Photomorphogenesis 57
              1.   General 57
              2.   Chloroplast Formation 57
        B.   Photomorphogenesis of External Structures 59
              1.   Higher Plants and Ferns 59
              2.   Spherulation and Sporulation of *Physarum* 60
              3.   *Trichoderma* Conidiation 60
              4.   *Phycomyces* 62
        C.   Transcriptional vs. Translational Control 62

III.   Rapid Effects 62
        A.   Electrical Changes 62
              1.   Possible Mechanisms 63
              2.   Blue Light and Electrical Changes in Phototropism 63
              3.   Electrical Changes in Blue Light-Induced
                    Morphogenesis 64
        B.   Changes in Organization of the Cytoplasm 65

IV.   Concluding Remarks 66

References 67

I. INTRODUCTION

Development and morphogenesis are among some of the least understood processes controlled by blue light. The excitation of some morphogenetic photoreceptors by a pulse of blue light forms a stable product. The product is so stable in some cases that the photoinductive effect can be remembered for a month when an organism is photoinduced in the cold.[1] The number of photons needed to trigger typical responses to blue light is so small that some kind of amplification must take place. It is thus worthwhile to obtain as many details as possible about the amplification steps, at the same time that the photoreceptors are being identified from the multitude of possible chromophores (see Chapter 1).

A. Early Effects and the Transduction Chain

The concept of early change is subjective and depends on the time scale of the responding organism (Figure 1). Some processes require long periods of continuous light to potentiate an effect. Tobacco cells in culture need days to form functional chloroplasts after transfer to continuous blue light.[2] In other systems, pulses are sufficient: stabilization of *Trichoderma* aerial hyphae leading to sporulation begins between 2 and 3 hr after a brief pulse of blue light[3,4] (Figure 1b). In attempts to find a causal chain, one must define early effects as the first photoinduced effects detected occurring before a new structure becomes evident. In many experiments, early changes overlap with late ones that occur during structural change. Clearly as techniques become more sophisticated, and events closer to the photoact are discerned, the findings discussed in this chapter will be superannuated.

B. Biochemical Development

Blue-light treatments often lead to the development of pigments or secondary metabolites. Such development is not strictly morphogenesis, defined as the appearance of new morphological structures. When biochemical development is induced by light, the response can sometimes be traced to specific enzymes, giving a useful handle for molecular studies. Biochemical development of single products is thus often, rightly or wrongly, used as a model for morphogenesis. We discuss a few examples here for comparison. Carotenoids in fungi apparently primarily serve as photoprotectants.[5] Two alternative mechanisms may be considered to explain the accumulation of carotenoids caused by illumination: photoactivation of preformed enzymes or *de novo* synthesis of the enzyme molecules. There is now clear evidence for regulation of mRNA levels in *Neurospora*; four of the new in vitro translated polypeptides appear as early as 10 min from the onset of light.[6]

UV-B light (290 nm most effective) controls sporulation in certain fungal systems.[7] UV-B also induces flavonoid photoprotectants in parsley cell suspension cultures (290 nm most effective). The synthesis of the mRNAs and subsequently the enzymes of the phenylpropanoid pathways leading to flavonoids are photoinduced. This was shown by blot hybridizations using total cellular RNA or RNA synthesized in vitro by isolated nuclei and cloned cDNAs specific for some of the mRNAs involved. The induction by UV-B light gives rise to transient increases in the rates of transcription of the genes encoding the respective enzymes.[8,9] (Figure 2).

Such definitive identification of the level of control is difficult in the widely used photomorphogenetic systems where the gene products acting in early stages are unknown. Nevertheless, considerable attention has been focused on transcription of mRNAs and their translation to proteins, because the appearance of new structures implies new gene products. This has been shown when morphogenesis was induced by factors other than light.

C. Transcription/Translation and Morphogenesis without Blue Light

The slime mold *Dictyostelium discoideum* has been especially amenable to such studies.[10]

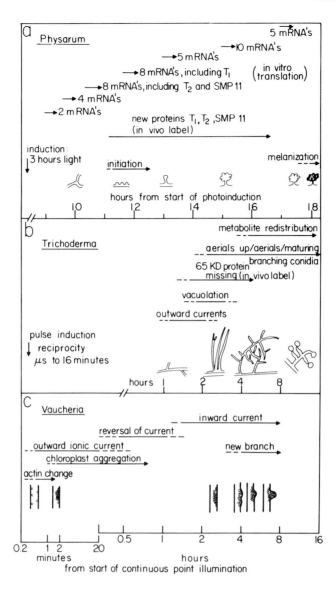

FIGURE 1. Time courses of morphogenesis and effects preceding or accompanying new structures. (a) Sporulation of *Physarum*. T_1 and T_2 are tubulins, SMP11 is a sporangiophore-specific protein specific to the induced state which appears when mRNA is translated in vitro. Arrows indicate new mRNA activities. Many of these species appear transiently,[37] appearing and disappearing sequentially. The onset of visible morphogenesis is between 10.5 and 11 hr after the start of photoinduction. Data condensed from Putzer et al.[37] (b) Sporulation (conidiation) of *Trichoderma*. Aerial hyphae of dark control colonies grow upward until about 2 hr after their initiation near the growing perimeter. They then fall over and join the horizontal mycelial mat. Aerial hyphae of photoinduced colonies are stabilized, so that a vertically growing ring clearly marks the position of the perimeter when the colony received light.[4] These as well as newly formed aerials eventually differentiate to conidiophores. Outward-flowing ionic currents,[58] an increase in vacuolation.[3] and the disappearance of a major pulse-labeled polypeptide spot on gels (see Figure 3) all precede the first signs of morphogenesis. There is enough time between light and the earliest known changes for many reactions to occur. (c) Induction of a new side branch by local irradiation in *Vaucheria*. A new branch begins to appear about 4 hr after the start of irradiation (continuous light). Note the double time scale: a change in organization of actin microfilaments[69,70] and a blue light-specific outward ionic current[56] are the earliest changes, followed closely by aggregation of chloroplasts. If these changes are all part of the transduction chain, it will be interesting to see how electrical and cytoskeletal changes are related to initiation of a new branch. Developmental scheme drawn from photographs of Weisenseel and Kicherer.[42]

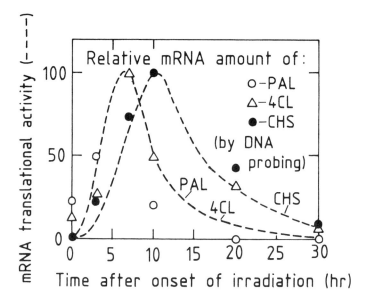

FIGURE 2. Photoinduction of mRNAs in parsley cell cultures. The amount of mRNA assayed by hybridization to cloned DNA probes for three enzymes of the phenylpropanoid pathway closely followed kinetics of mRNAs assayed by in vitro translation. PAL, phenylalanine ammonia lyase; 4CL, 4-coumarate:CoA ligase; CHS, chalcone synthase. These studies were done with UV (290 nm most effective) light, and photoinduction leads to protective compounds, not true morphogenesis. Such biochemical models might closely resemble the early stages of morphogenetic pathways or, perhaps, other kinds of information may be needed to determine new structures. (Redrawn from Kuhn, D. N. et al., *Proc. Natl. Acad. Sci. U.S.A.*, 81, 1102, 1984. With permission.)

Much evidence has also accumulated for sporulation and germination of fungi.[11,12] In discussing blue light, we cannot be blind to the other end of the visible spectrum. Red light (acting through phytochrome, as inferred by red, far-red reversibility alone) modulates transcription of specific genes[13] and overall transcriptional activity of isolated nuclei,[14] especially in higher plant systems.

The ability to follow specific gene products by cloning and hybridization can help to resolve controversies from overinterpretation of earlier data.[15,16] Evidence that the transcription and/or translation of particular mRNAs are turned on or off by light is needed. The most used inhibitors (puromycin, cycloheximide, and actinomycin D) can indirectly block turnover as well as syntheses,[17] leading to problems in interpreting experiments using inhibitors in conjuction with blue light.[18]

The earliest appearing mRNA species may still be too minor in quantity to detect. Even so, we could imagine that programs of gene expression will eventually be catalogued. We will then be able to ask what steps occur between photoproduct and gene expression.

D. Rapid Effects

Jaffe[19] proposed electrical controls, suggesting that morphogenesis begins from the outside, with gene expression modulated only quite late in the chain. This argument could be applied to enzymatic mechanisms as well. Some of the rapid effects of blue light (see Section III) occur within minutes. An example would be phototropism and the light-growth response of *Phycomyces*. Blue light causes a rapid decrease in growth rate with a 1 min lag in etiolated cucumber and sunflower,[20] and with a lag of less than 5 min in dark-adapted cucumber seedlings.[21] In such sensory chains, growth and movement reorient, but no novel structures appear. In these rapid responses, there is no need to postulate changes in gene expression, although it is difficult to preclude them. *Phycomyces* is an organism with both rapid and morphogenetic response chains. Besides phototropism, light induces sporangiophore initiation. There are mutants (*madC, D, E, F, G*) whose defects interfere with bending but not sporangiophore initiation, but *madA* and *madB* are blocked in both pathways. This suggests that the pathways are distinct, with a common pigment system input. The mutants may help show how much of each pathway is shared, and whether early effects are necessary for control of gene expression. Organisms where morphogenesis can be induced by stress as well as light also help dissect the pathway (e.g., *Trichoderma*[3]). Blue light may act as a transient stress,[3,7,18] and its effects can be replaced by a transient stress that has helped assay for genetic lesions in photoreception.[22] Two operational models for morphogenesis which are not mutually exclusive are that (1) new specific molecules (e.g., inducers, second messengers, mRNAs), or (2) new physical properties (e.g., electric fields and currents) could program the appearance of new structures.

II. TRANSCRIPTION AND TRANSLATION

A considerable mass of data has accumulated based on a variety of methods that points to transcriptional or translational controls in responses to blue light (see reviews[7,18,23,24]). Examples of the kinds of experiments being done are summarized and compared in Table 1. A wide range of questions are being asked and in many cases the organisms and methods are ideal for only certain aspects of the overall problem while omitting others.[18] In the following discussion we will attempt to summarize the progress.

A. Cellular (Internal) Photomorphogenesis
1. General
Light can control the morphogenetic development of structures, especially organelles within cells. Chloroplast development, either naturally during germination or the greening of etiolated tissue, is the best-studied case. Biochemical and other aspects of greening are discussed in detail in Volume 1, Chapter 7.

2. Chloroplast Formation
The steps leading from rudimentary proplastids or from etioplasts to functioning chloroplasts in higher and lower plants and *Euglena* include both biochemical development of defined products and morphogenesis of new membranes and complexes. Such development is often blocked when the chloroplast pigments are absent. Some of the massively produced products probably help organize the new structures (see Volume 1, Chapter 7). A drawback of greening systems is the multitude of pigments present that may be possibly required both as products and controlling elements (protochlorophyll, chlorophylls, cryptochrome(s), and phytochrome). Much of the experimentation on plastid development has been done with white light, and some of the products that are induced (light-harvesting complexes, photosynthetic apparatus) may in turn act as photoreceptors for further development even before measureable photosynthesis starts.

Blue light is the best stimulator of both cytoplasmic and plastid rRNAs in *Euglena*. Dark-grown *Euglena* cells transferred to a resting medium undergo only one further division. They remain viable and undergo normal chloroplast development when illuminated. Blue light is most effective in inducing regeneration of phototransformable protochlorophyll(ide).[25] The conditions were somewhat artificial but may have allowed visualization of the role of each photoacceptor controlling greening and separate controls by different photoreceptors. Red was tenfold less active than blue, so the receptor cannot be a typical cryptochrome. Previous

Table 1
COMPARISON OF STUDIES ON TRANSCRIPTIONAL AND TRANSLATIONAL CHANGES INDUCED BY BLUE LIGHT

Organism	Response type	First visible (hr)	Type of light	Methods used	Assay	Ref.
Euglena	Greening	72	Cont.	Silver-stained 2-d gel	Steady-state levels of proteins	28
	Greening	4	Cont.	RNA hybridized to plastid DNA restriction fragments	Plastid mRNA changes	29
Spirodela	Greening	3	Cont.	Gel of plastid RNAs	Rapidly synthesized new mRNA	74
Acetabularia	Morphogenesis, enzyme induction	24	Cont.	Enucleation, in- hibitors, density labeling	Photostimulation of enzyme	41
Tobacco cells	Greening	24—48	Cont.	RNA hybridized to cloned plas- tid genes	mRNA	2, 24
Physarum	Inhibition of spherulation	16—32	Cont.	2-d Gels, in vivo and in vitro label	Inhibition at mRNA and protein level	36
	Induction of sporulation	9	Pulse	In vivo, in vitro label; 2-d gels, RNA hybrid- ized to cloned genes	Proteins and mRNA	37
Trichoderma	Induction of sporulation	2	Pulse	In vivo label, 2-d gel	Proteins	Fig. 3
Phycomyces	Diameter and adaptation kinet- ics of sporangiophores	96	Pulse or cont.	In vivo label, 2-d gels	Flavoprotein photo- receptor candidates	39
Neurospora	Carotenogenesis	0.17	Pulse	In vivo, in vitro label, 2-d gels	Proteins, mRNA mRNA	6 8
Parsley cells	Flavonoid synthesis	3	Pulse	RNA hybridized to cloned genes in vivo, in vitro label 2-d gels	Proteins, comparison of kinetics	9

Note: The list is by no means complete; we attempt to draw comparisons between the different methods, time scales and organisms chosen. Some examples that do not strictly belong to morphogenesis have been included, as the methods and conclusions are often similar.

action spectra[26] showed that both a noncryptochromal blue light receptor and protochlorophyll(ide) control chlorophyll formation. There is also genetic evidence in *Euglena* for distinct blue and red/blue receptors.[27] The nongreen mutant W_3BUL lacks the entire plastid (i.e., photosynthetic pigments and genome), yet blue light still induces cytoplasmic rRNA biosynthesis. This is strong genetic evidence that there is a blue receptor independent of the plastid. Still, the situation is probably not that simple: in another mutant, Y_9ZNa1L, protochlorophyllide is lacking, but plastid DNA is apparently retained.[27] Some plastid enzymes that are thought to be nuclear encoded and cytoplasmically translated are repressed in dark-grown wild type cells. These are induced by light (red and blue). As expected, they cannot be induced in W_3BUL. In Y_9ZNa1L they are inducible, but only blue light is effective. From the data with these two mutants, it was inferred that wild-type *Euglena* has coregulation by at least two pigment systems.[27]

Monroy and Schwartzbach[28] compared the patterns on two-dimensional gels of total protein from dark-grown cells with those obtained after 72 hr of white light. Almost a quarter of the 650 polypeptides found were light induced, but the effects of different wavelengths were not separated. It would certainly be interesting to see changes earlier than 72 hr, especially in view of evidence[29] from hybridization that some plastid transcription is modified within 4 hr after transfer to light.

Mutants of *Chlorella* and *Scenedesmus,* algae which normally do not need light to green, require cryptochromal blue light action. Continuous blue light induces detectable greening of *Scenedesmus* C-2A′ within 4 hr. Here as in *Euglena*, special conditions were very helpful in unraveling the effects of different pigment systems. Greening later than 12 hr after light also depended on the newly formed photosynthetic apparatus.[30]

In most higher plants, red light is very effective in greening; i.e., removing a lag in synthesis which normally occurs when white light is used.

The greening of tobacco cell cultures (in contrast to tobacco plants) is under nearly exclusive blue light control. Dark-grown cells develop functional chloroplasts if transferred to constant blue light,[2] but red light is ineffective. Relatively high irradiances of blue light are required: about 10 $W \cdot m^{-2}$ continuous wide-band blue light, and the time scale is days. According to the subjective time scale of these cells, though, the changes studied could still be called early. The increase in the levels of the large subunit of ribulose bisphosphate carboxylase (RuBPCase) and the precursor of the 32-kDa membrane protein correlated with an increase in the mRNAs for these proteins. The kinetics after transfer to light were followed by dot hybridization to cloned DNA sequences.[2] This nicely supplements older work showing that the synthesis of the mRNA in *Spirodela* for the 32 kDa protein was primarily under blue light control.[31] This approach was less successful with the small subunit of RuBPCase which is nuclear encoded.[32] It is tempting to speculate that plastid and nuclear genes might be separately controlled at different levels by different pigment systems. The questions of whether blue light affects one or more cryptochromes, and extent of involvement of other blue light receptors, whether the receptors control both cytoplasmic and nuclear genomes, and whether second messengers are involved, are all intriguing, and need further research.

Inhibitors, such as diuron to block photosynthetic electron flow, help separate different photoreception chains (review[18]). Mutants can also be helpful but they are easier to isolate in microorganisms like *Euglena* (see above) than in higher plant cell lines. Complete action spectra on processes needing days of continuous light are rather cumbersome to obtain. Irradiance response curves at a few wavelengths could differentiate between cryptochromal and noncryptochromal chromophores. For cryptochrome, red light should be virtually ineffective; 455 nm should be somewhat more effective than 480 nm, and 530 nm should be orders of magnitude less effective than 480 nm. Such experiments should be only slightly more difficult than the work usually done with wide-band filters. Induction of the 32-kDa protein in *Spirodela* was shown in this way to be not typically cryptochromal.[31]

B. Photomorphogenesis of External Structures

The first biochemical products in morphogenesis of external structures are unknown. The earliest biochemical handles usually studied are structural components such as cell wall polysaccharides or proteins or spore pigments, and these may be very far indeed from the initial photoreactions.

1. Higher Plants and Ferns

In higher plants, calculations must be made to sort out cryptochrome from phytochrome effects.[33] There is good evidence for distinct cryptochrome control in anthocyanin synthesis (see Volume 1, Chapter 7). Continuous blue light regulates leaf growth, size, and shape.[34] The kinetics are slow, and the molecular studies done for greening have not yet been attempted here.

Red light maintains filamentous protonemal growth of gametophytes of several ferns, while continous blue light induces biplanar growth (review[35]). Microbeam experiments suggest that blue light acts at the nucleus, but the new molecular genetic techniques to follow specific gene products have not been applied.[35] Electrophysiological work has also been done with ferns (see Figure 5a), and the system is promising, though genetic analyses are much harder than with fungi.

2. Spherulation and Sporulation of Physarum

The plasmodial slime mold *Physarum* has two developmental pathways leading to resting structures which are determined by light and by the growth conditions. Spherulation is induced by starvation, but not by light. Continuous blue light inhibits this transition of starving plasmodia growing in liquid suspension to spherules. A 3-hr exposure to blue or red light induces formation of specialized fruiting bodies (sporulation) on starving plasmodia growing on solid substrates. The photoreceptor for sporulation is not a typical cryptochrome; it is not known whether red and blue light are sensed by the same chromophore.

No spherulation-related changes in in vivo labeling of proteins were found by two-dimensional electrophoresis at 8 hr after transfer to starvation medium. About 16 to 32 hr later, 20 differentiation-specific proteins appeared, along with 26 new in vitro translatable mRNA species.[36] These changes are not found in continuous blue light.

Sporulation is more appealing for study as it is specifically induced by light (rather than inhibited). New gene products must play some role in sporulating plasmodia. Putzer et al.[37] used in vivo ^{35}S-methionine labeling and two-dimensional gels to follow the steps leading to sporulation. Starving plasmodia were exposed to a 3-hr photoinduction, a short pulse on a subjective time scale for development (Figure 1a). The first changes were detected when enhanced labeling began at 11 hr after photoinduction, when there were already the first visible signs of morphogenesis. Isolation and in vitro translation of mRNA provided a sharper time resolution and allowed detection of three times more new protein species than when the labeling was in vivo. The earliest detectable in vitro translated spots appeared at 9 hr, 2 hr before visible morphogenesis. Two new spots appearing on the in vivo and in vitro gels (T1 and T2) were identified as tubulins. Another major new spot, SMP11, has not yet been characterized, but must be a quantitatively important cell component at this stage, as it could be detected by Coomassie blue staining as well as by radioactive labeling. Northern blot hybridization using a cDNA probe for the SMP11 mRNA showed that this mRNA was light induced.

Many translatable mRNAs also disappeared during sporulation. One of these was the message for plasmodial actin. Nuclei isolated from sporulating plasmodia still continued to produce actin-specific sequences when tested in vitro. Actin expression thus seems to be regulated via mRNA processing or degradation, rather than at the level of transcription.[38]

3. Trichoderma Conidiation

Conidiation of *Trichoderma* is induced by a brief pulse of blue light with a cryptochrome action spectrum.[3] The research up to 1980 on relationships between conidiation and transcription with this fungus has been reviewed.[18] We expected that photoinduction might result in the appearance of novel gene products; the time scale of *Trichoderma* (Figure 1b) would allow studies following a blue light pulse of seconds. *Trichoderma* photoresponse (including likely photoreceptor) mutants are available, but nothing is known about the early photo-controlled gene products, in contrast to *Physarum* where at least actin and tubulins have been identified. We recently found that labeling of a major spot near 65 kDa and pH 6.8 disappeared within 2 hr after a 4-min blue light pulse (Figure 3b). Samples labeled from the time of photoinduction to 30 min afterwards gave patterns similar to the dark control. Most of the spots detectable at various times thereafter were the same for dark controls and

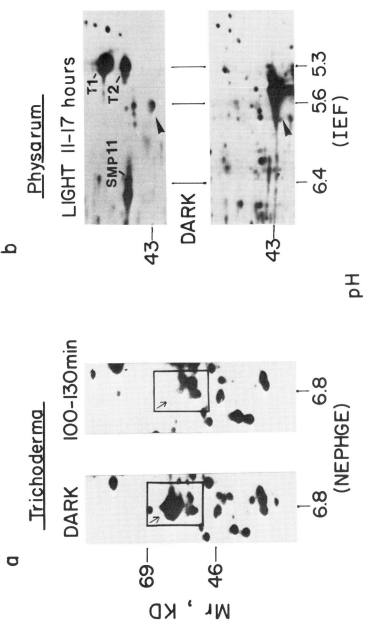

FIGURE 3.   Two-dimensional electrophoresis to map early changes in protein synthesis following a light-pulse-inducing sporulation. (a) Regions of gels of a 10,000 *xg* supernatant of extracts from *Trichoderma* colonies labeled in vivo from 100 to 130 min after a photoinductive blue light pulse did not show a major spot present in the dark control gels (Horwitz, B. A., Marks, R., and Gressel, J., previously unpublished data). First dimension, nonequilibrium pH gradient electrophoresis (NEPHGE), second dimension, SDS-PAGE. (b) New proteins induced by light in *Physarum*. Labeling was from 11 to 17 hr after light, when morphogenesis had already begun. Gels of total cell proteins; first dimension, isoelectric focusing; second dimension, SDS-PAGE. (Data presentation modified from Putzer et al.[36] Original print kindly supplied by Dr. T. Schreckenbach, Max-Planck Institute for Biology, Martinsried, West Germany.)

photoinduced colonies, in contrast to *Physarum* where a considerable fraction of the poly-peptide spots changed. The reason for this may be that the mycelial front of an induced *Trichoderma* colony continues to advance, requiring that vegetative characteristics must be retained.

4. Phycomyces

Many well-studied *Phycomyces* mutants are available. A high-speed pelletable fraction of sporangiophores enriched for plasma-membrane ATPase has recently been studied using two-dimensional gels of in vivo steady-state labeled proteins.[39] Two spots, probably flavoproteins as they were labeled with both ^{14}C-riboflavin labeled and ^{35}S sulfate, changed in intensity with the growth conditions: both spots were darker in light- than dark-grown material, and their ratio changed. These proteins are candidates for *Phycomyces* crypto-chromes. Photocontrol at the protein level may be related to adaptation of the photoreceptive apparatus, as dark- and light-grown cultures both produced sporangiophores. Blue light also induces the appearance of sporangiophores in vegetative *Phycomyces* cultures; there is ap-parently still no research correlating this process with transcription and translation.

C. Transcriptional vs. Translational Control

In much of the preceding discussion, translational and transcriptional controls were dis-cussed together. The rates of transcription, translation, and degradation together determine the level of a given protein. There may clearly be separations between transcription and translation with an interval between the time when a mRNA is transcribed to when it is actually translated. A day between transcription and translation has been found for some proteins during phytochrome-induced fern spore germination.[40] In vitro translation allows a more direct assay of the levels of the mRNAs; as cloned DNA probes and isolated nuclear in vitro translation systems become available, the assays should become much more specific. These methods have already been applied, particularly in *Physarum,* where the control of actin was not transcriptional, assuming that isolated nuclei behave as they do in vivo.

The huge unicellular alga *Acetabularia* offers a unique opportunity to separate nuclear from cytoplasmic control: large anucleate cell fragments are easily detached by cutting off the rhizoid containing the sole nucleus. Continuous blue light stimulates, via cryptochrome, growth and morphogenesis (hair whorl formation), both in the presence of actinomycin D and in enucleated cells. Studies on UDP-glucose pyrophosphorylase (Figure 4) clearly showed that blue-light induction does not require the nucleus.[41] Transcription and translation by the cytoplasmic genomes could not be excluded by enucleation experiments, but organellar activity was excluded by using chloramphenicol to block organelle transcription. It will be interesting to ascertain whether UDP-glucose pyrophosphorylase is directly related to mor-phogenesis. Another question is whether nonnuclear control is general, or is a special adaptation of this remarkable alga.

III. RAPID EFFECTS

There are some early effects of blue light that could be explained without invoking gene expression models. There are rapid modulations of growth (see Section I of this chapter). We will discuss two other aspects here: electrical changes and mechanical (cytoskeletal) modifications. Other possibilities, such as enzyme activation/inhibition and nonelectrical membrane effects are just as worthy of consideration at present, and are the subjects of Chapters 5 and 6.

A. Electrical Changes

Bioelectric changes are rapid and are appealing as possible early transduction steps by

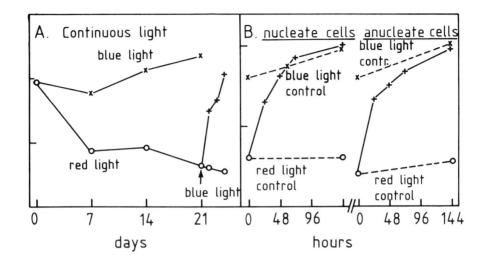

FIGURE 4. Regulation of an enzyme by blue light without participation of the nucleus in *Acetabularia*. (A) UDP-glucose pyrophosphorylase (an enzyme involved in sucrose synthesis) activity in extracts of cells grown under continuous red or blue light, and stimulation by blue light. (B) Blue light stimulation (solid line) with or without nucleus (see text). Dashed lines, blue and red light controls as in (A). Enucleated cells of this alga still respond morphogenetically to blue light, strengthening the conclusions from inhibitor studies. The regulatory effect of blue light does not require the nucleus. UDP-glucose pyrophosphorylase might be regulated in a similar way to gene products that control morphogenesis, or might even be part of the transduction chain. (Redrawn from Schmid, R., in *Blue Light Effects in Biological Systems.* Senger, H., Ed., Springer-Verlag, Berlin, 1984, 428. With permission.)

analogy to nerve systems. Electric current patterns (or the local membrane permeabilities of which these are external manifestations) might transduce physical-chemical changes which provide new information sufficient to determine new structures.[19,42]

1. Possible Mechanisms

Exogenous electric fields can mediate changes in cell shape and orientation, perhaps through interaction with the cytoskeleton[43,44] and are even used to enhance the rate of bone healing.[19] Constant electric fields have been found to orient hyphal growth in *Schizophyllum*.[45] Excitation by light involves a momentary separation of charge; this can be converted into macroscopic separation across a membrane following redox reactions, as occurs in photosynthetic membranes. In microorganisms such as the cyanobacterium *Phormidium*, photomovement seems to be controlled by electrical gradients set up by photosynthetically active light[46] (see also Volume 1, Chapter 11).

How electric fields/currents induce a new gene product has yet to be shown; i.e., the situation is similar to that of light. The developmental significance of light-mediated electrical responses in plants has recently been reviewed by Racusen and Galston.[47] Interpretation of phytochrome-mediated electrical changes in green plants is complicated by the electrical activity of photosynthetic pigments.[48] Blue light-controlled responses are thus advantageous for electrical as well as for molecular studies on early effects. Blue light is especially useful when it can be shown that red light is ineffective or when roots, albino mutants, or fungi are used.

2. Blue Light and Electrical Changes in Photoptropism

Forty years ago Schrank[49] found that a phototropic stimulus to *Avena* coleoptiles sets up a large surface potential difference between the two sides, which precedes the redistribution of growth. One would have to show how the potential difference is related to distribution of chemical effectors and in turn to which the latter bind to make useful causal inferences.

Johnsson[50] found that auxin is needed for the potential change. The electrical assymetry may well be the expression of auxin redistribution, leading us back to ask how light initiated the chain reaction. Racusen and Galston[51] recently found that a red light pretreatment changes the sensitivity of coleoptile cell membrane potentials to blue light. In *Phycomyces,* where correlations could be tested with mutants, the evidence for electrical field changes is tenuous. Mogus and Wolken[52] reported potential changes lasting 2 to 10 sec and occurring about 2.5 min after illumination. The potential difference was measured across the ends of a detached sporangiophore, and may have been completely unphysiological. Groves and Gamow[53] used less destructive techniques and were not able to find such a response. These studies concern movement more than morphogenesis.

3. Electrical Changes in Blue Light-Induced Morphogenesis

Electric potential differences and currents are measured across the plasmalemma of single cells, at the surface of aerial organs, or in a growth medium with low but measurable conductivity. Some of the many experiments are illustrated in Figure 5. Bean hook opening involves irreversible changes in structure. Near UV (NUV) and UV irradiations (365 and 254 nm) caused large surface potential changes in etiolated bean hypocotyl hooks[54] (Figure 5b); the NUV, but not the UV, response required the presence of carotenoids. Blue light with a cryptochrome action spectrum was also effective.[55]

Perhaps the clearest correlation of blue light, electrical changes, and morphogenesis was found in the coenocytic alga *Vaucheria.*[56] A blue light-induced outward ionic current preceded and was precisely correlated with the aggregation of chloroplasts. The initiation of a new side branch followed chloroplast aggregation in response to point irradiations without blue light (Figure 1C). Mutants with plastid aggregation or with current defects were not available to check whether these early electrical effects are essential for branching, but the correlation is very suggestive. Further aspects of the *Vaucheria* response are discussed in the next section.

Turian[57] found that hyphal tips of *Neurospora* and *Allomyces* were able to chemically reduce indicator dyes such as methylene blue, while subapical regions were not. This gradient in reducing power disappeared at the onset of conidiation (not photoinduced). This was not analyzed electrophysiologically but such a finding should suggest a change in ionic currents preceding conidiation. With this in mind, rapid electric current pattern modifications were measured after photoinducing *Trichoderma* mycelia. There were outward currents specific to the photoinduced state, but no rapid modification of the currents near the growing tips.[58] There is a good correlation[59] between growth rate and inward currents in *Achlya.* The failure to detect a change in the current entering the growing *Trichoderma* hyphal tips was then actually not surprising; the growth rate of the tips did not change after photoinduction. Outward currents appeared in the region of the hyphae formed 90 to 120 min after the short photoinductive pulse (Figure 5d). This is much too late to be the first transduction step, but the currents may be related to a stressed state of the photoinduced mycelia. Localized current loops on a scale of 30 μm or less can be inferred from these measurements (measuring positions A, B, C, and D in Figure 5d). The membranes of the induced hyphae may have become leaky to ions, in small discrete patches. The significance of this for morphogenesis is not clear yet. Light also increased vacuolation of *Trichoderma* hyphae,[3] known to be a starvation (stress) response in *Trichoderma.*[60] Blue light stimulated membrane potential changes in *Neurospora* at irradiances (20 W · m^{-2} for 20 to 25 min), relatively high for this system.[61] This was not correlated with morphogenesis, though that should now be possible with the recent studies on photoinduction of perithecia formation.[62]

A possible point for morphogenetic regulation might be alternate cyanide insensitive respiration. The branch point from the usual respiratory pathway is through a flavoprotein,[63] a good candidate for blue light photoreception. Interestingly, salicylhydroxamic acid (SHAM),

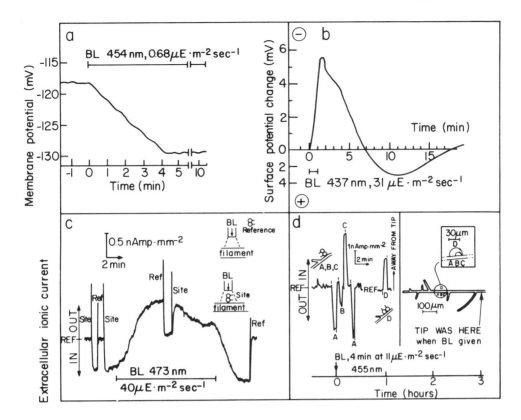

FIGURE 5. Examples of electrical changes induced by blue light. The amount of light used (photon exposure) and the time course of each response are indicated. (a) Change in membrane potential recorded intracellularly in apical cells of the gametophyte of the fern *Onoclea sensibilis* (redrawn from Cook, T. J., Racusen, R. H., and Briggs, W. R., *Planta,* 159, 300, 1983; with permission). (b) Surface potential change recorded on hypocotyl hooks of bean seedlings (redrawn from Hartmann, E. and Schmidt, K., in *The Blue Light Syndrome,* Senger, H., Ed., Springer-Verlag, Berlin, 1980, 221; with permission). (c) Outward current induced by microbeam irradiation of a *Vaucheria sessilis* filament. Current measured with an extracellular vibrating electrode. The sketches in the upper-right corner indicate the orientation and direction of vibration of the probe at the measuring site and the reference site (REF) (redrawn from Blatt, M. R., Weisenseel, M. H., and Haupt, W., *Planta,* 152, 513, 1981; with permission). (d) Localized outward currents specific to photoinduced *Trichoderma* hyphae. Currents recorded with an extracellular vibrating electrode; the small sketches indicate different measuring sites; REF indicates the signal at reference site far from the hyphae (redrawn from Horwitz, B. A. et al., *Plant Physiol.,* 74, 912, 1984; with permission). No outward currents were detected near dark control hyphae; most outward currents were detected 60 to 150 min after photoinduction. These currents did not correlate with any obvious structural feature, and might be correlated with the stressed[18] state of photoinduced hyphae.

which blocks the terminal oxidation step of this pathway, inhibited a rapid electrical response to blue light.[64] SHAM did not inhibit phototropism in corn coleoptiles,[65] which could easily be explained by a lack of inhibitor penetration.

B. Changes in Organization of the Cytoplasm

Fifty years ago, Bottelier[66] reported that light decreased cytoplasmic streaming in oat coleoptile cells within minutes, and showed that the effective wave band was in the blue. The action spectrum, within the resolution then obtainable, corresponded to that of phototropism (Figure 6). Blue light also decreased the viscosity of the cytoplasm in *Elodea* leaves.[67] The response was quantified by measuring the degree of displacement of organelles after centrifugation of the leaves. The action spectra measured were typically cryptochromal. The absolute threshold for the response was similar to that for oat coleoptile phototropism. For

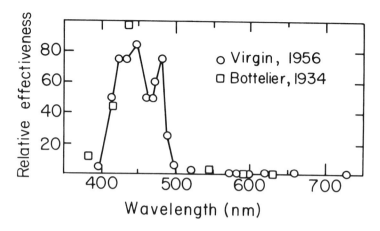

FIGURE 6. Rapid effects of blue light on mechanical properties of the cytoplasm. Action spectra for decreased viscosity of the cytoplasm of *Elodea* leaf cells[67] and for a decrease in the rate of cytoplasmic streaming in oat coleoptile cells.[66] The short time between illumination and the onset of these effects suggests a rather direct effect of cryptochrome photoreception on cytoskeletal proteins. Even minutes, though, could allow a second messenger to act.

further discussion and action spectra, see Seitz.[68] These effects do not seem to belong to any morphogenetic pathway.

A fibrillar network forms in *Vaucheria*, in response to irradiation with blue light.[69] It is postulated that the chloroplasts aggregate because they become trapped in it. In dark-adapted filaments, thick parallel cortical fibers can be observed with Nomarski optics. Within 30 sec of the onset of blue light, the thick fibers seem to become destabilized, forming a finer, more net-like pattern. Withing 60 sec, the first chloroplasts become trapped.[69] Data from the use of inhibitors phalloidin and Cytochalasin B, as well as electron microscopy, suggest that these fibers have actin-like properties.[70]

IV. CONCLUDING REMARKS

There is ample evidence for changes at the mRNA and protein level following blue light irradiation. These are not surprising, as new structures or activities are often induced. There is still a gap of at least an hour between illumination in pulse-induced systems and the first measured changes in proteins or mRNA levels or syntheses. This is sufficient time to allow many intermediary but more primary processes to occur. Further work will be needed to show whether this limitation is due to the mechanism of the response or to the time scales chosen by the investigators. Very little is known about early events, but blue light is an easily manipulatable inducer and many mutants are available or easy to isolate in blue light-responding microorganisms. In the much discussed membrane model for phytochrome, the main difficulty is not in detecting rapid effects, but in showing that they are related to the transduction chain leading to morphogenesis.[71] Mutants are unavailable in most such organisms. A question which may soon be addressable is whether electrical and/or mechanical changes are primary steps occurring as immediate consequences of irradiation, or whether these result from transcriptional, translational, or enzyme activity changes. The multitude of enzymes (see Chapter 5) that could be modified by blue light certainly leaves this open for speculation. Still, the reported irradiances required for those changes in vitro are many orders of magnitude greater than those needed to induce many photomorphogenetic changes.

A promising nondestructive method for studying metabolism in vivo is NMR. This may be easier to correlate with the response than in vitro assays. ^{31}P NMR has recently been applied to blue light-induced greening in *Scenedesmus*.[72] Light induced changes in the levels of some phosphate compounds within 5 min of the onset of illumination, as rapid as many electrical changes. It will be interesting to see more quantitative and more complete kinetics using NMR.

Transcriptional changes could be much more rapid than usually expected or studied. They occur within a few minutes following introduction of metabolic inducers to bacteria. Whether transcriptional changes coupled to translational changes can occur sufficiently rapidly and to a great enough extent to account for rapid photo- or hormone-induced growth effects is an avidly discussed point at present.

Neurospora can be transformed with cloned DNA sequences.[73] Such methods may be applicable to other blue light-sensitive fungi as well. Ideally, cloned genes could be used to complement mutants blocked in specific stages of morphogenesis. A problem is that the treatments required for transformation completely disrupt normal development. This means that the newly introduced gene would be present throughout development, where it could have multiple effects.

Finally, we must ascertain the general validity of conclusions reached using blue light *vis a vis* larger questions of morphogenesis. Blue light, in many cases, may act as a stress,[7,18] suggesting action at key metabolic branch points. We hope that blue light-induced photomorphogenesis may be an excellent model for many morphogenetic processes, and conclusions from the kind of experiments we have discussed here will not be limited to plants or fungi.

REFERENCES

1. **Gressel, J. and Galun, E.,** Sporulation in *Trichoderma:* a model system with analogies to flowering, in *Cellular and Molecular Aspects of Floral Induction,* Bernier, G., Ed., Longman, London, 1970, 152.
2. **Richter, G.,** Blue light control of two plastid mRNA's in cultured plant cells, *Plant Mol. Biol.,* 3, 271, 1984.
3. **Horwitz, B. A., Gressel, J., and Malkin, S.,** The quest for *Trichoderma* cryptochrome, in *Blue Light Effects in Biological Systems,* Senger, H., Ed., Springer-Verlag, Berlin, 1984, 237.
4. **Gressel, J. and Galun, E.,** Morphogenesis in *Trichoderma:* photoinduction and RNA, *Dev. Biol.,* 15, 575, 1967.
5. **Krinksy, N. I.,** The protective function of carotenoid pigments, in *Photophysiology,* Vol. III, Giese, A. C., Ed., Academic Press, New York, 1968, 123.
6. **Mitzka-Schnabel, U., Warm, E., and Rau, W.,** Light-induced changes in the protein pattern translated *in vivo* and *in vitro* accompanying carotenogenesis in *Neurospora crassa* and *Fusarium aquaeductuum,* in *Blue Light Effects in Biological Systems,* Senger, H., Ed., Springer-Verlag, Berlin, 1984, 264.
7. **Gressel, J. and Rau, W.,** Photocontrol of fungal development, in *Encyclopedia of Plant Physiology, New Series,* Shropshire, W. Jr. and Mohr, H., Springer-Verlag, Berlin, 1983, 603.
8. **Kreuzalar, F., Ragg, H., Fautz, E., Kuhn, D. N., and Hahlbrock, K.,** UV-induction of chalcone synthase mRNA in cell suspension cultures of *Petroselinum hortense, Proc. Natl. Acad. Sci. U.S.A.,* 80, 2591, 1983.
9. **Kuhn, D. N., Chappell, J., Boudet, A., and Hahlbrock, K.,** Induction of phenylalanine ammonia-lyase and 4-coumarate: CoA ligase mRNAs in cultured plant cells by UV light or fungal elicitor, *Proc. Natl. Acad. Sci. U.S.A.,* 81, 1102, 1984.
10. **Barklis, E. and Lodish, H. F.,** Regulation of *Dictyostelium discoideum* mRNAs specific for prespore or prestalk cells, *Cell,* 32, 1139, 1983.
11. **Brambl, R., Dunkle, L. D., and van Etten, J. L.,** Nucleic acid and protein synthesis during fungal spore germination, in *The Filamentous Fungi,* Vol. 3, Smith, J. E. and Berry, D. R., Eds., John Wiley & Sons, New York, 1978, 94.
12. **Lovett, J. S.,** Macromolecular synthesis in *Blastocladiella,* in *Fungal Differentiation,* Smith, J. E., Ed., Marcel Dekker, New York, 1983, 211.
13. **Silverthorne, J. and Tobin, E. M.,** Demonstration of transcriptional regulation of specific genes by phytochrome action, *Proc. Natl. Acad. Sci. U.S.A.,* 81, 1112, 1984.

14. **Mösinger, E. and Schäfer, E.,** *In vivo* phytochrome control of *in vitro* transcription rates in isolated nuclei from oat seedlings, *Planta,* 161, 444, 1984.

15. **Attridge, T. H. and Smith, H.,** Density-labelling evidence for the blue light-mediated activation of phenylalanine ammonia lyase in *Cucumis sativa* seedlings, *Biochem. Biophys. Acta,* 343, 452, 1974.

16. **Schopfer, P.,** Phytochrome control of enzymes, *Annu. Rev. Plant Physiol.,* 28, 223, 1977.

17. **Schimke, R. T. and Doyle, D.,** Control of enzyme levels in animal tissues, *Annu. Rev. Biochem.,* 39, 929, 1970.

18. **Gressel, J.,** Blue light and transcription, in *The Blue Light Syndrome,* Senger, H., Ed., Springer-Verlag, Berlin, 1980, 133.

19. **Jaffe, L. F.,** Developmental currents, voltages and gradients, in *Developmental Order: Its Origin and Regulation,* Alan R. Liss, New York, 1982, 183.

20. **Cosgrove, D. J.,** Rapid suppression of growth by blue light. Occurrence, time course, and general characteristics, *Plant Physiol.,* 67, 584, 1981.

21. **Gaba, V. and Black, M.,** Photocontrol of hypocotyl elongation in de-etiolated *Cucumis sativus* L. Rapid responses to blue light, *Photochem. Photobiol.,* 38, 469, 1983.

22. **Horwitz, B. A., Gressel, J., Malkin, S., and Epel, B. L.,** Modified cryptochrome *in vivo* absorption in *Trichoderma dim* photosporulation mutants, *Proc. Natl. Acad. Sci. U.S.A.,* 82, 2736, 1985.

23. **Senger, H. and Briggs, W. R.,** The blue light receptor(s): primary reactions and subsequent metabolic changes, in *Photochemical and Photobiological Reviews,* Smith, K. C., Ed., Plenum Press, New York, 1981, 1.

24. **Richter, G.,** Blue light effects on the level of translation and transcription, in *Blue Light Effects in Biological Systems,* Senger, H., Ed., Springer-Verlag, Berlin, 1984, 253.

25. **Alhadeff, M., Coronado, R., Figueroa, N., and Schiff, J.,** Regulation of photochlorophyll(ide) levels in dark-grown non-dividing *Euglena* — control by light, *Photochem. Photobiol.,* 38, 679, 1983.

26. **Egan, J. M., Dorsky, D., and Schiff, J. A.,** Events surrounding the early development of *Euglena* chloroplasts. VI. Action spectra for the formation of chlorophyll, lag elimination in chlorophyll synthesis, and appearance of TPN-dependent triose phosphate dehydrogenase and alkaline DNAse activities, *Plant Physiol.,* 56, 318, 1975.

27. **Schiff, J. A. and Schwartzbach, S. D.,** Photocontrol of chloroplast development in *Euglena,* in *Biology of Euglena,* Vol. 3, *Physiology,* Buetow, D. W., Ed., Academic Press, New York, 1982, 313.

28. **Monroy, A. F. and Schwartzbach, S. D.,** Photocontrol of the polypeptide composition of *Euglena.* Analysis by two-dimensional gel electrophoresis, *Planta,* 158, 249, 1983.

29. **Chelm, B. K., Hallick, R. B., and Gray, P. W.,** Transcription program of the chloroplast genome of *Euglena gracilis* during chloroplast development, *Proc. Natl. Acad. Sci. U.S.A.,* 76, 2258, 1979.

30. **Senger, H. and Bishop, N. I.,** The development of structure and function in chloroplasts of greening mutants of *Scenedesmus.* I. Formation of chlorophyll, *Plant Cell Physiol.,* 13, 633, 1972.

31. **Gressel, J.,** Light requirements for the enhanced synthesis of a plastid mRNA during *Spirodela* greening, *Photochem. Photobiol.,* 27, 167, 1978.

32. **Richter, G., Hundrieser, J., Gross, M., Schultz, S., Bottlander, K., and Schneider, C.,** Control of gene expression in blue light-dependent chloroplast differentiation, in *Advances in Photosynthesis Research,* Vol. IV, Sybesma, C., Ed., Martinus Nijhoff/Dr. W. Junk, The Hague, 1984, 853.

33. **Schäfer, E. and Haupt, W.,** Blue-light effects in phytochrome mediated responses, in *Encyclopedia of Plant Physiology, New Series,* Vol. 16B, Shropshire, W., Jr. and Mohr, H., Eds., Springer-Verlag, Berlin, 1983, 723.

34. **Beggs, C. J., Barth, J., and Schäfer, E.,** The effect of light on cotyledon and primary leaf growth in white light grown *Sinapis alba, Physiol. Plant.,* 57, 114, 1983.

35. **Furuya, M.,** Photomorphogenesis in ferns, in *Encyclopedia of Plant Physiology, New Series,* Vol. 16B, Shropshire, W., Jr., and Mohr, H., Eds., Springer-Verlag, Berlin, 1983, 569.

36. **Putzer, H., Werenskiold, K., Verfuerth, C., and Schreckenbach, T.,** Blue light inhibits slime mold differentiation at the mRNA level, *EMBO J.,* 2, 261, 1983.

37. **Putzer, H., Verfuerth, C., Claviez, M., and Schreckenbach, T.,** Photomorphogenesis in *Physarum:* induction of tubulins, sporulation-specific proteins and their mRNA's, *Proc. Natl. Acad. Sci. U.S.A.,* 81, 7117, 1984.

38. **Schreckenbach, T.,** personal communication, 1985.

39. **Pollock, J. A., Lipson, E. D., and Sullivan, D. T.,** Analysis of flavoproteins in *Phycomyces* sporangiophores: candidates for the blue light photoreceptor, *Planta,* 163, 506, 1985.

40. **Zilberstein, A., Gressel, J., Arzee, T., and Edelman, M.,** Early morphogenetic changes during phytochrome-induced fern germination. II. Transcriptional and translational events, *Z. Pflanzenphysiol.,* 114, 109, 1984.

41. **Schmid, R.,** Blue light effect on morphogenesis and metabolism in *Acetabularia,* in *Blue Light Effects in Biological Systems,* Senger, H., Ed., Springer-Verlag, Berlin, 1984, 419.

42. **Weisenseel, M. H. and Kicherer, R. M.,** Ionic currents as control mechanism in cytomorphogenesis, in *Cell Biol. Monographs, Cytomorphogenesis in Plants,* Vol. 8, Kiermayer, O., Ed., Springer-Verlag, Berlin, 1981, 381.

43. **Cooper, M. S. and Keller, R. E.,** Perpendicular orientation and directional migration of amphibian neural crest cells in dc electrical fields, *Proc. Natl. Acad. Sci. U.S.A.,* 81, 160, 1984.

44. **Luther, P. W., Peng, H. B., and Lin, J. J.-C.,** Changes in cell shape and actin distribution induced by constant electric fields, *Nature (London),* 303, 61, 1983.

45. **De Vries, S. C. and Wessels, J. G. H.,** Polarized outgrowth of hyphae by constant electrical fields during reversion of *Schizophyllum commune* protoplasts, *Exp. Mycol.,* 6, 95, 1982.

46. **Nultsch, W. and Häder, D.-P.,** Photomovement of motile microorganisms, *Photochem. Photobiol.,* 29, 423, 1979.

47. **Racusen, R. H. and Galston, A. W.,** Developmental significance of light-mediated electrical responses in plant tissue, in *Encyclopedia of Plant Physiology, New Series,* Vol. 16B, Shropshire, W., Jr. and Mohr, H., Eds., Springer-Verlag, Berlin, 1983, 687.

48. **Montavon, M., Horwitz, B. A., and Greppin, H.,** Enhancement by red light of far-red stimulated intracellular potential changes in spinach leaf mesophyll cells, *Plant Physiol.,* 73, 671, 1983.

49. **Schrank, A. R.,** Note on the effect of unilateral illumination on the transverse electrical polarity in the *Avena* coleoptile, *Plant Physiol.,* 21, 362, 1946.

50. **Johnsson, A.,** Photoinduced lateral potentials in *Zea mays, Physiol. Plant.,* 18, 574, 1965.

51. **Racusen, R. H. and Galston, A. W.,** Phytochrome modifies blue light-induced electrical changes in corn coleoptiles, *Plant Physiol.,* 66, 534, 1980.

52. **Mogus, M. A. and Wolken, J. J.,** *Phycomyces:* electrical response to light stimuli, *Plant Physiol.,* 53, 512, 1974.

53. **Groves, P. M. and Gamow, R. I.,** Intracellular recordings from *Phycomyces, Plant Physiol.,* 55, 946, 1975.

54. **Hartmann, E. and Schmidt, K.,** Effects of UV and blue light on the biopotential changes in etiolated hypocotyl hooks of dwarf beans, in *The Blue Light Syndrome,* Senger, H., Ed., Springer-Verlag, Berlin, 1980, 221.

55. **Hartmann, E.,** Influence of light on the bioelectric potential of the bean *(Phaseolus vulgaris)* hypocotyl hook, *Physiol. Plant.,* 33, 266, 1975.

56. **Blatt, M. R., Weisenseel, M. H., and Haupt, W.,** A light-dependent current associated with chloroplast aggregation in the alga *Vaucheria sessilis, Planta,* 152, 513, 1981.

57. **Turian, G.,** Reducing power of hyphal tips and vegetable apical dominance in fungi, *Experientia,* 32, 989, 1976.

58. **Horwitz, B. A., Weisenseel, M. H., Dorn, A., and Gressel, J.,** Electric currents around growing *Trichoderma* hyphae, before and after photoinduction of conidiation, *Plant Physiol.,* 74, 912, 1984.

59. **Armbruster, B. L. and Weisenseel, M. H.,** Ionic currents traverse growing hyphae and sporangia of the mycelial water mold *Achlya debaryana, Protoplasma,* 115, 65, 1983.

60. **Pitt, D. E. and Bull, A. T.,** Free amino acid pools of *Trichoderma aureoviride* during conditions of glucose-limited growth and glucose starvation, *Arch. Microbiol.,* 130, 180, 1981.

61. **Potapova, T. V., Levina, N. N., Belozerskaya, T. A., Kritsky, M. S., and Chailakhian, L. M.,** Investigation of electrophysiological responses of *Neurospora crassa* to blue light, *Arch. Microbiol.,* 137, 262, 1984.

62. **Degli-Innocenti, F., Pohl, U., and Russo, V. E. A.,** Photoinduction of protoperithecia in *Neurospora crassa* by blue light, *Photochem. Photobiol.,* 37, 49, 1983.

63. **Day, D. A., Arron, G. P., and Laties, G. G.,** Nature and control of respiratory pathways in plants: the interaction of cyanide-resistant respiration with the cyanide-sensitive pathway, in *The Biochemistry of Plants,* Vol. 2, Davies, D. D., Ed., Academic Press, New York, 1980, 198.

64. **Cooke, T. J., Racusen, R. H., and Briggs, W. R.,** Initial events in the tip-swelling response of the filamentous gametophyte of *Onoclea sensibilis* L. to blue light, *Planta,* 159, 300, 1983.

65. **Vierstra, R. D. and Poff, K. L.,** Mechanism of specific inhibition of phototropism by phenylacetic acid in corn seedlings, *Plant Physiol.,* 67, 1011, 1981.

66. **Bottelier, H. P.,** Über der Einfluss ausseren Faktoren auf die Protoplasmastromung in der *Avena* Koleoptile, *Rec. Trav. Bot. Neerl.,* 31, 474, 1934.

67. **Virgin, H.,** An action spectrum for the light-induced changes in viscosity of plant protoplasm, *Physiol. Plant.,* 5, 575, 1952.

68. **Seitz, K.,** Cytoplasmic streaming and cyclosis of chloroplasts, in *Encyclopedia of Plant Physiology, New Series,* Vol. 7, Haupt, W. and Feinleib, M. E., Eds., Springer-Verlag, Berlin, 1979, 150.

69. **Blatt, M. R. and Briggs, W. R.,** Blue light-induced cortical fiber reticulation concomittant with chloroplast aggregation in the alga *Vaucheria sessilis,,* *Planta,* 147, 355, 1980.

70. **Blatt, M. R., Wessels, N., and Briggs, W. R.,** Actin and cortical fiber reticulation in the siphonaceous alga *Vaucheria sessilis, Planta,* 147, 363, 1980.

71. **Quail, P. H.,** Rapid action of phytochrome in photomorphogenesis, in *Encyclopedia of Plant Physiology, New Series, Photomorphogenesis,* Vol. 16A, Shropshire, W., Jr., and Mohr, H., Eds., Springer-Verlag, Berlin, 1983, 178.

72. **Oh-Hama, T. and Ruyters, G.,** ^{31}P-NMR studies in *Scenedesmus* C-2A′ in darkness and blue light, in *Blue Light Effects in Biological Systems,* Senger, H., Ed., Springer-Verlag, Berlin, 1984, 323.

73. **Dhawale, S. S., Paietta, J. V., and Marzluf, G. A.,** A new, rapid and efficient transformation procedure for *Neurospora, Curr. Genet.,* 8, 77, 1984.

74. **Rosner, A., Jacob, K. M., Gressel, J., and Sagher, D.,** The early synthesis and possible function of a 0.5×10^6 mRNA after transfer of dark-grown *Spirodela* plants to light, *Biochem. Biophys. Res. Commun.,* 67, 383, 1975.

Chapter 5

CONTROL OF ENZYME CAPACITY AND ENZYME ACTIVITY

Günter Ruyters

TABLE OF CONTENTS

I.   Introduction 72

II.   General Aspects 72
        A.   The Range of Enzymes Affected by Blue Light 72
        B.   Mechanisms of Enzyme Control 72

III.   Control Mechanisms Realized in Blue Light 76
        A.   Coarse Control by Blue Light 76
        B.   Fine Control by Blue Light 77
                1.   Direct Action of Blue Light 77
                2.   Indirect Action of Blue Light 78

IV.   Conclusion 83

Acknowledgments 84

References 84

I. INTRODUCTION

In plants, blue light (BL) — absorbed by the so-called BL photoreceptor(s) — exerts important regulatory effects on many processes including movement, morphogenesis, cell growth and cell cycle, stomatal opening, and metabolism.[1-3] Probably the most important action of BL is that on metabolism, since all processes in plant life ultimately are based on one or more metabolic reactions. Carbon as well as nitrogen metabolism is affected by light quality; so, for example, carbohydrate degradation, protein and RNA synthesis, respiration and nonphotosynthetic CO_2 fixation have been found to be stimulated by BL; the underlying mechanism(s), however, are only poorly understood.[4-7] Since metabolic reactions are catalyzed by enzymes, these are certainly probable candidates to be regulated by light quality and should therefore play an important role within the sequence light perception — transduction — metabolic response. In trying to analyze the BL action on metabolism, several studies indeed have focused on the investigation of enzymes in recent years.

II. GENERAL ASPECTS

A. The Range of Enzymes Affected by Blue Light

During the last 20 years a large number of enzymes of various organisms have been reported to be affected by BL. Table 1 tries to summarize the present information; in the studies cited, always the organisms, not the isolated enzymes, have been irradiated. BL effects mediated via phytochrome or chlorophyll should not be included in this list, although the involvement of only the BL photoreceptor(s) has not been demonstrated in many cases. In fact, respective wavelength dependences exist only for pyruvate kinase, glutamate synthase (GOGAT), and 2-oxoglutarate dehydrogenase of the chlorophyll-free *Chlorella* mutant no. 20.[18,29,56]

The large number of enzymes listed in Table 1 can be divided mainly into two groups, photosynthetic enzymes including those of pigment synthesis and photorespiration, and carbohydrate-degrading enzymes,[57] indicating a special role of BL for these processes. Indeed the regulation of the development and maintenance of a functioning photosynthetic apparatus is being discussed as one of the most important physiological functions of BL in green algae and higher plants, whereby — at least in the early stages of development — the necessary carbon and energy are supplied from the degradation of endogenous carbohydrate reserves.[4,7,58-61,100]

B. Mechanisms of Enzyme Control

After demonstrating that quite a number of enzymes have been reported to be under BL control, the regulatory mechanism(s) involved should be discussed. At first the question should be answered how enzymes can be regulated in general; in a second step we will then deal with the problem, which of the possibilities at our disposal are realized in the case of BL-mediated regulation of enzymes.

Control of enzyme activity and, thus, metabolic control can be achieved mainly in two ways (Figure 1). In so-called "coarse control" the amount of active enzyme, the enzyme capacity,[62] is regulated mostly by changing the rate of synthesis or degradation; regulation of transcription and/or of translation may be involved, with substrate induction and product repression of enzymes being well-known examples of this mechanism. In some other cases, changes in enzyme capacity are caused by release of the enzyme from bound or inactive forms (e.g., β-amylase in germinating cereals) by processing of proenzymes (e.g., the small subunit of RubP carboxylase) or by production of inactivating proteins (e.g., for acid invertase). Regardless of the respective mechanism, it is a characteristic feature for coarse control that these changes in enzyme capacity occur rather slowly within hours or even days,

Table 1
ENZYMES REPORTED TO BE UNDER BLUE LIGHT CONTROL

Enzyme	Plant	Effect	Ref.
ALA dehydratase	*Chlorella* mutant no. 20	+	8
	Scenedesmus C-2A'	+	9
ALA synthetase	*Scenedesmus* C-2A'	+	9
Alcohol dehydrogenase	*Phycomyces blakesleeanus*	−	10
Aldolase	*Euglena gracilis*	+	11
c-AMP phosphodiesterase	*Neurospora crassa*	+	12
	Phycomyces blakesleeanus	+	13
Catalase	*Triticum aestivum*	+	14
Chitin synthetase	*Phycomyces*	+	97
DOVA-dehydrogenase ⎱ DOVA-transaminase ⎰	*Scenedesmus* C-2A'	+	9
Fumarase	*Euglena gracilis*	+	15
Glucose-6-phosphate dehydrogenase	*Acrochaetium daviesii*	+	16
	Wolffia arrhiza	+	17
Glutamate pyruvate transaminase (GPT)	*Chlorella* mutant no. 20	−	18
Glutamate synthase (GOGAT)	*Chlorella* mutant no. 20	+	18
Glyceraldehyde phosphate dehydrogenase (NAD-dep.)	*Chlorella* mutant no. 20	−	8
	Chlorogonium elongatum	+	19
	Scenedesmus obliquus	+	20
Glyceraldehyde phosphate dehydrogenase (NADP-dep.)	*Chlamydomonas reinhardii*	+	21
	Chlorella mutant no. 20	+	22
	Chlorella vulgaris	+	23
	Chlorogonium elongatum	+	19
	Euglena gracilis	+	11
	Scenedesmus obliquus	−	20
	Vicia faba	+	24
	Zea mays	+	24
Glycolate oxidase	*Dolichos lablab*	−	25
	Pisum	+	26
	Triticum aestivum	+	14
	Phaseolus	+	27
	Zea mays	+	27
Glyoxylate aminotransferase	*Phaseolus*	+	27
	Zea mays	+	27
Hydroxypyruvate reductase	*Pharbitis nil*	+	28
	Triticum aestivum	+	14
Isocitratase	*Chlorogonium elongatum*	−	19
α-Ketoglutarate dehydrogenase	*Chlorella* mutant no. 20	−	29
NAD kinase	*Neurospora crassa*	+	30
NADP protochlorophyllide oxidoreductase	*Nicotiana tabacum* (callus)	−	31
Nitrate reductase	*Agmenellum quadruplicatum*	−	32
	Chlamydomonas reinhardii	+	33
	Pisum arvense	+	34
	Zea mays	+	34
	Sinapis alba	+	35
	Triticum	+	36
	Zea mays	+	37
Nitrite reductase	*Chlorella pyrenoidosa*	+	38
Phenylalanine ammonialyase	*Cucumis sativus*	+	39
	Zea mays	+	40
Phosphoenolpyruvate carboxylase	*Chlamydomonas reinhardii*	+	21
	Chlorella ellipsoidaea	+	41
	Chlorella mutant no. 20	+	8, 42
	Chlorella mutant no. 125	+	43

Table 1 (continued)
ENZYMES REPORTED TO BE UNDER BLUE LIGHT CONTROL

Enzyme	Plant	Effect	Ref.
	Chlorogonium elongatum	+	19
	Dolichos lablab	+	25
	Scenedesmus obliquus	+	20
	Vicia faba	+	24
	Zea mays	+	24
6-Phosphogluconate dehydrogenase	*Acrochaetium daviesii*	+	16
Phosphoglycerate kinase	*Euglena gracilis*	+	11
Pyruvate kinase	*Acetabularia mediterranea*	+	44
	Chlorella mutant no. 20	+	45
	Chlorella mutant no. 125	+	Ruyters (unpublished)
	Pisum sativum	+	46
	Scenedesmus C-2A'	+	47
Ribulose bis-phosphate carboxylase	*Anabaena flos-aquae*	−	48
	Chlamydomonas reinhardii	+	21
	Chlorella mutant no. 20	+	8
	Chlorogonium elongatum	+	19
	Cyanidium caldarium	+	98
	Euglena gracilis	+	11
	Hordeum	+	49
	Microcystis aeruginosa	−	48
	Nicotiana tabacum (callus)	+	50
	Vicia faba	+	24
	Zea mays	+	24
RNA polymerase	*Hordeum*	+	49
Succinate dehydrogenase	*Euglena gracilis*	+	15
UDPG-pyrophosphorylase	*Acetabularia mediterranea*	+	44, 51

Note: The complex enzyme systems of carotenoid biosynthesis are not quoted.[52-55,99] Further BL-affected enzymes of the chlorophyll-free *Chlorella* mutant no. 20 are shown in Table 4 and Figure 2.

Modified from Ruyters, G., in *Blue Light Effects in Biological Systems,* Senger, H., Ed., Springer-Verlag, Berlin, 1984. With permission.

so that they are suitable only for the long-term regulation of developmental processes. They can be measured — and this is very important — as changes in the maximal velocity V_{max} of an enzymatic reaction, assayed under optimal in vitro conditions.

"Fine control", on the other hand, regulates the activity of preexisting enzyme molecules without affecting their amount. This can be achieved by changing the concentration of substrates, products, and cofactors (including ions) as well as of effectors, usually metabolites, which either isosterically or allosterically increase or decrease the activity of enzyme molecules present. Competitive inhibition (e.g., of succinate dehydrogenase by malonate) or allosteric feedback inhibition (e.g., of phosphofructokinase by PEP) are well-known examples for this latter mechanism. Enzyme activity can also be modulated by posttranslational events such as chemical modification (phosphorylation - dephosphorylation, e.g., of phosphofructokinase or pyruvate kinase; reduction - oxidation, e.g., of several enzymes of the Calvin cycle) or physical interactions (dissociation - association, e.g., of glyceraldehyde 3-phosphate dehydrogenase). Since no protein synthesis is involved, fine control can be achieved rather quickly, so that it is suitable for the rapid fine adjustment of metabolic pathways.

As already mentioned, measurement of enzyme activity under optimal in vitro conditions can only indicate the amount of active enzyme, the enzyme capacity which, however, is not (necessarily) equivalent to the in vivo activity. Changes in the latter are very difficult

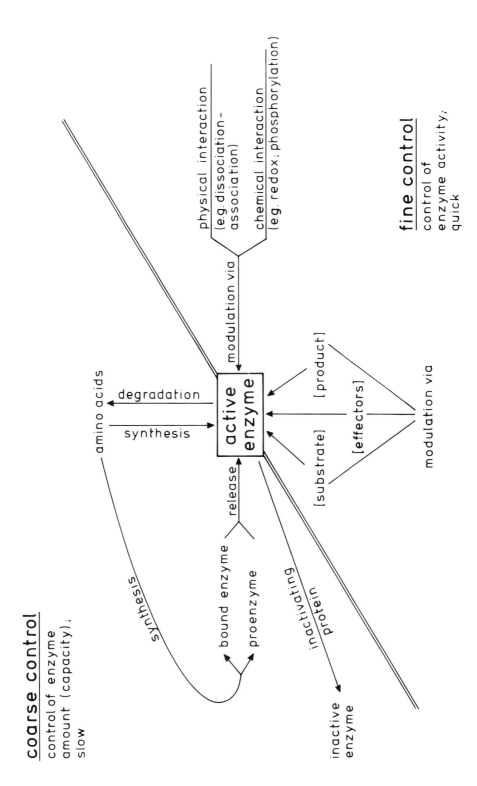

FIGURE 1.   Control mechanisms of enzyme capacity and enzyme activity.

to assess, since the enzyme activity would have to be determined under in vivo conditions, i.e., at in vivo concentrations of substrate, cofactors, effectors, ions, enzyme protein itself, and at physiological pH and temperature. Since plant metabolism obviously has the ability to respond quickly and successfully to variations of environmental conditions, fast changes in enzyme activity certainly have to occur. Therefore and in spite of the difficulties discussed, fine control of metabolic pathways is being investigated with emphasis.

The respective attempts have already led to some general concepts. The key to the analysis of fine control is the fact that most of the control of a pathway is concentrated in only a few steps, the regulatory reactions. These reactions are far from thermodynamic equilibrium, since they are catalyzed by enzymes controlled by factors other than substrate concentration. Certainly a change in flux through a sequence will involve an alteration in the activity of each enzyme — no reaction through which there is a net flux can be at equilibrium — but many enzymes are so fast that their reactions in vivo do not show much deviation from equilibrium; these enzymes are regarded as being unable to regulate flux directly.

The experimental approach to the study of fine control involves the following steps.[63] First the nonequilibrium reactions of the pathway under discussion must be identified. Comparison of the enzyme capacities of all enzymes within the sequence gives preliminary indications, since regulatory enzymes usually have low capacities. A more reliable method is the comparison of the apparent equilibrium constants of the enzymes with the ratio of their products and substrates present in vivo, the mass action ratio: substantial difference (20-fold or more[64]) is clear evidence that a reaction is displaced from equilibrium in vivo. This permits the reaction to control flux, but does not prove that it does.[65,66] Proof that a nonequilibrium reaction is regulatory may be obtained by demonstrating that the amount of substrate present in vivo changes in opposite direction to flux through the pathway when the latter is varied.

If an enzyme proves to be regulatory, it is necessary to study its properties in vitro in order to establish how it might be controlled in vivo. These studies will probably disclose some more attributes, usually associated with regulatory enzymes. Their substrate dependences often deviate from Michaelis-Menten kinetics, as shown by nonlinear Lineweaver-Burk plots; this indicates cooperativity, i.e., the enzyme activity is allosterically regulated. Besides this, many reactions catalyzed by regulatory enzymes have high free energy changes and are practically irreversible. Often regulatory enzymes are located near branchpoints in metabolic pathways.

Based on all information obtained from these investigations, one can then try to construct a hypothesis that explains how the respective pathway might be regulated in vivo. Finally, this hypothesis should be proved by showing that variation in the proposed effectors regulates flow through the pathway in the manner required by the hypothesis.

These theoretical considerations make it obvious that information about fine control is even more difficult and laborious to obtain than about coarse control. Therefore, although fine control of metabolic pathways appears to be understood for the most part, data for particular environmental conditions are lacking to great an extent.

III. CONTROL MECHANISMS REALIZED IN BLUE LIGHT

A. Coarse Control by Blue Light

After this introduction of mechanisms at the disposal of plant cells for regulating metabolism via enzymes, the question is, which of these mechanisms are realized in BL — a question which is not easy to answer. In most studies enzyme activity has been measured with an in vitro assay under optimized conditions, thereby obtaining the maximal velocity V_{max} of that enzymatic reaction. As already stated, V_{max} is a measure for the amount of active enzyme, the enzyme capacity. BL-mediated changes in V_{max} are considered to be indicative of changes in the rate of protein synthesis or degradation, and the data of Table 1 have been interpreted this way by the respective authors. In some cases supporting experiments with protein synthesis inhibitors have been reported.[43,45] Both results together strongly suggest the involvement of protein synthesis in the BL control of enzymes, but proof can come only from somewhat more sophisticated techniques like immunological procedures or comparative density labeling. In fact, de novo synthesis of RubPCase could be proved in cultured tobacco cells by demonstrating an increase in the level of translatable mRNA for its small subunit; this was achieved by isolating the respective in vitro translation products by immunoprecipitation.[50] Very recently, Richter[67,101] was also able to show photoregulation of two mRNAs, which are encoded in the chloroplast DNA. First, the level of mRNA for the large subunit of RubPCase appeared to be high in blue light-irradiated cells and in their plastids, respectively, but almost undetectable in dark-grown cells, indicating a coordinate regulation of the two subunits. Second, the concentration of the mRNA coding the precursor polypeptide of the 32-kDa protein was enhanced drastically by BL in tobacco cells. In the future, these new techniques will certainly prove that BL also induces the synthesis of other enzymes, for which changes in V_{max} or inhibitor studies as yet are the only indications of such a BL action. Since the capacity of several enzymes is shown to decrease in BL (see Tables 1 and 4), light-stimulated degradation of enzyme proteins must also be taken into account. Detailed analysis, however, has not been performed yet. In summary, these results demonstrate that BL is able to influence enzyme synthesis (and degradation?) — in other words, can exert coarse control.

B. Fine Control by Blue Light

In discussing BL-mediated fine control of metabolism, i.e., modulation of the activity of already existing enzyme molecules, two possibilities should be distinguished: direct and indirect action of BL.

1. Direct Action of Blue Light

In the case of direct action, the blue radiation has to be absorbed by the enzyme molecule itself or by its prosthetic group. Not whole organisms, but the isolated and more-or-less purified enzymes, were irradiated in these experiments. Since most enzymes and their coenzymes do not absorb visible radiation, this mechanism is restricted to very few cases. The example studied best is probably nitrate reductase; containing FAD, cytochrome b_{557} and molybdenum as cofactors, the inactivated enzyme from fungi, green algae, and higher plants can be reactivated by BL, indicating that FAD is the primary photoreceptor.[68,69,102]

Some other enzymes have been reported to become BL-sensitive upon the addition of flavin mononucleotide (FMN) in vitro (Table 2). In most cases, the enzyme activity is reduced by BL in the presence of FMN. Only the activity of glycine oxidase of the colorless *Chlorella* mutant no. 125 is enhanced by BL under such conditions, and the activity of a FAD containing glucose oxidase — suppressed by removing FAD — can be restored with blue light in the presence of FMN. The physiological significance of these results, however, is controversial.

Recently, the influence of BL on enzymes in vitro in the presence of FMN was again investigated in connection with studies on enzymes of the carbohydrate metabolism of *Chlorella*[103] (Table 3). The activities of all enzymes tested were reduced by BL under such conditions, which is in sharp contrast to results of experiments in which enzyme capacities were measured in crude extracts from blue-light-irradiated intact cells of the *Chlorella* mutant no. 20. Phosphofructokinase was found to be not affected, and pyruvate kinase was found to be stimulated by BL under these conditions.[45,77]

These contrary findings raise new doubts about the physiological meaning of the effect of BL on enzymes in the presence of FMN in vitro.

Table 2
BLUE LIGHT EFFECTS ON ENZYME IN VITRO UPON ADDITION OF FMN(I)

Enzyme	Material	Effect	Ref.
Amino acid oxidase	Hog kidney	−	70
Glucose oxidase	*Aspergillus*	+	71
Glycine oxidase	*Chlorella* mutant no. 125	+	72
Glycolate oxidase	*Nicotiana*	−	73
Lactate dehydrogenase	Yeast	−	70
Malate dehydrogenase	*Euglena*	−	74
	Nicotiana	−	74
	Zea mays	−	74
	Pig heart	−	74
Nitrate reductase	*Spinacia*	−	75
Ribulose bisphosphate carboxylase	*Anabaena*	−	48
	Microcystis	−	48
	Scenedesmus	−	48
	Sphaeocystis	−	48
Transketolase	*Nicotiana*	−	76
Uricase	Hog liver	−	70
Xanthine oxidase	Bovine milk	−	70

After Ruyters, G., in *Blue Light Effects in Biological Systems*, Senger, H., Ed., Springer-Verlag, Berlin, 1984. With permission.

Table 3
BLUE LIGHT EFFECTS ON ENZYMES IN VITRO UPON ADDITION OF FMN(II)

Enzyme	Material	Effect
Glyceraldehyde phosphate dehydrogenase (NAD-dep.)	Rabbit muscle	—
Hexokinase	Yeast	—
Phosphofructokinase	Rabbit muscle	—
	Chlorella mutant no. 20	—
Pyruvate kinase	Rabbit muscle	—
	Chlorella mutant no. 20	—

2. Indirect Action of Blue Light

Indirect action of BL in this context means that the blue radiation is not absorbed by the enzyme itself or its prosthetic group and thus does not act directly upon the enzyme molecule; it rather means that BL — after being absorbed by the BL photoreceptor(s) — affects the enzyme activity indirectly by changing the concentration of substrates, products and cofactors (including ions) or that of activating or inhibiting effectors. Also chemical or physical modification of enzymes must be taken into account as possible regulatory mechanisms. As already mentioned, reliable investigations about this type of metabolic fine control are rather difficult and lacking in BL research. Since BL is known to stimulate pathways such as respiration within only a few minutes,[78] fine control of enzyme activity is certainly involved in regulating flux through pathways and should therefore be studied with emphasis.

The regulation of carbohydrate breakdown and respiration of green algae, especially of *Chlorella* and its chlorophyll-free mutant no. 20, is currently under our investigation according to the strategy outlined above.[4,5,57] The resulting picture is still rather incomplete, but it might serve as an example for what has to be done. Nearly all enzymes of carbohydrate breakdown and many of closely related pathways have been investigated in the yellow *Chlorella* mutant no. 20 (Figure 2, Table 4). Stress was laid upon the glycolytic pathway, since carbohydrate degradation is also stimulated by blue light under anaerobic conditions.[81] Among the glycolytic enzymes measured, low enzyme capacities were found for hexokinase, phosphofructokinase, and pyruvate kinase,[45,77] indicating that these might have a regulatory function, a result which is consistent with the general view of the regulation of glycolysis.[82]

In the following, pyruvate kinase (PK) of *Chlorella* mutant no. 20 was studied in detail. First, the mass action ratio for the PK reaction was calculated from the data of Kowallik and Scheil[83] for ADP and ATP and from our own preliminary results for phosphoenolpyruvate (PEP) and pyruvate; it gives values of 0.10 in the case of dark-kept and 0.58 in the case of BL-irradiated cells. These data allow the following conclusions. First, the mass action ratio for PK of the yellow *Chlorella* mutant is in both conditions far from the equilibrium constant of 11,000;[65] this supports the above assumption that the enzyme plays a regulatory role in *Chlorella* mutant no. 20, too. Second, the difference in the mass action ratios between dark- and BL-kept cells, arising from an increase in pyruvate and ATP and a decrease in PEP and ADP, respectively, indicates that indeed the in vivo activity of PK is stimulated by BL.

After proving that PK plays a regulatory role also in the yellow *Chlorella* mutant no. 20, kinetic and regulatory properties of this enzyme were investigated in vitro. Our further studies revealed that the alga contains two isoenzymes of PK.[84] As concluded from comparable experiments with greening pea plants, they are possibly located in the cytosol and the plastid, respectively.[46] The ratio of the two isoenzymes of *Chlorella* mutant no. 20 is influenced by BL, which seems to convert one isoenzyme into the other.[84,85] The mechanism of this possible interconversion has not been elucidated yet; it is well known, however, that PK of animal systems is subject to posttranslational phosphorylation-dephosphorylation processes.[86]

The isoenzymes of the *Chlorella* mutant cells then were partially purified in order to study the influence of several effectors. Table 5 shows the effect of AMP, ATP, P_i, citrate, F-6-P, and FbP on the maximal velocity V_{max}. While AMP, P_i, F-6-P, and FbP exert no marked influence, ATP and citrate inhibit V_{max} of both isoenzymes. Table 6 demonstrates the influence of the same effectors on $S_{0.5}$ (PEP); i.e., on substrate affinity. Again P_i, F-6-P, and FbP have only minor influence. ATP and citrate increase the $S_{0.5}$ values, i.e., they decrease the affinity; on the contrary, AMP decreases the $S_{0.5}$ values; i.e., increases the affinity of both isoenzymes towards PEP. In total, these two tables indicate that both PK isoenzymes have major potential for (allosteric) regulation in vitro. This may be of physiological significance as well: for *Chlorella* mutant no. 20, an enhanced ATP and decreased ADP and AMP level after BL irradiation were demonstrated;[83] altered levels of adenine nucleotides were also found for *Chlamydomonas*[87] and for *Scenedesmus* C-2A′.[88] These results indicate that BL might be able to control the in vivo activity of PK via changes in the concentration of substrate — ADP — and of effectors — AMP and ATP. At present nothing is known about the concentration of other important metabolites such as citrate. Altogether these results, obtained according to the concept outlined above, strongly suggest that the glycolytic pathway in *Chlorella* is, at least partially, regulated at the PK step.

Another enzymatic reaction, on which BL seems to exert fine control, is the step from PEP to oxaloacetate, catalyzed by PEP carboxylase (PEPCase). Evidence for BL-enhanced in vivo activity of this enzyme came from $^{14}CO_2$ fixation studies of Miyachi and co-workers.[6,41,43] They established that BL stimulates dark $^{14}CO_2$ fixation, much of the label immediately being found in aspartate derived via oxaloacetate from carboxylation of PEP. This result indicates that the in vivo PEPCase activity increases rapidly under blue light.

A new approach to the understanding of the fine control of carbohydrate degradation by blue light is currently being made by Oh-hama and Ruyters by using ^{31}P nuclear magnetic resonance spectroscopy (^{31}P NMR).[88,89,105] Since the glycolytic intermediates — except the

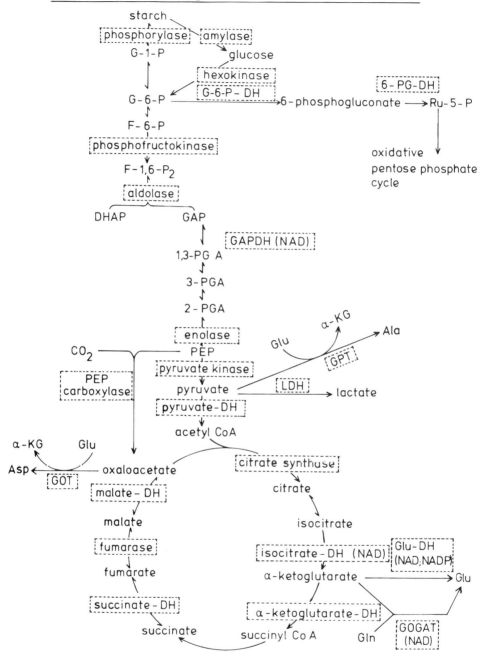

carbohydrate degradation in Chlorella M 20

FIGURE 2.   Pathways of carbohydrate degradation and of related amino acid metabolism in simplified form; blue light-affected enzymes of the chlorophyll-free *Chlorella* mutant no. 20 in frames. (After Ruyters, G., in *Blue Light Effects in Biological Systems,* Senger, H., Ed., Springer-Verlag, Berlin, 1984. With permission.)

Table 4
BLUE LIGHT EFFECTS ON ENZYMES OF CARBOHYDRATE BREAKDOWN AND RELATED PATHWAYS IN THE *CHLORELLA* MUTANT NO. 20

Enzyme	Enzyme capacity[a]	Effect[b]	Ref.
Amylase	70	(+)	79
Phosphorylase	15	0	79
G-6-P dehydrogenase	70	0	77
6-PG dehydrogenase	200	(−)	77
Hexokinase	140	(−)	77
Phosphofructokinase	80	0	77
Aldolase[c]	160	+	77
Glyceraldehyde phosphate dehydrogenase (NAD-dep.)	1400	(−)	8
Enolase	400	−	77
Pyruvate kinase	100	+	45
PEP carboxylase	20	+	42
Pyruvate dehydrogenase	10	?	103
Lactate dehydrogenase	30	(+)	103
Citrate synthase	300	(−)	80
Isocitrate dehydrogenase (NAD-dep.)	40	+	80
Isocitrate dehydrogenase (NADP-dep.)	80	(+)	80
α-Ketoglutarate dehydrogenase	30	−	29
Succinate dehydrogenase	15	−	103
Fumarase	800	0	80
Malate dehydrogenase	12000	(−)	80
Glutamate oxaloacetate transaminase (GOT)	1500	(−)	18
Glutamate pyruvate transaminase (GPT)	250	−	18
Glutamate dehydrogenase (NAD-dep.)	150	0	18
Glutamate dehydrogenase (NADP-dep.)	30	0	18
Glutamate synthase (GOGAT)	5	+	18
Glutamine synthetase (GS)	50	+	104

[a] Enzyme capacity given in nmol/min mg protein.
[b] Effects smaller than 15% after 24 hr of BL irradiation are indicated with brackets.
[c] Could not be distinguished between glycolytic and photosynthetic enyzme activity.

Modified after Ruyters, G., in *Blue Light Effects in Biological Systems*, Senger, H., Ed., Springer-Verlag, Berlin, 1984.

final product pyruvate — and many important co-factors and effectors are phosphate compounds, they are detectable by ^{31}P NMR. ^{31}P NMR has become a well-established method for the study of the bioenergetics and metabolism of bacterial and animal systems. Changes in the level of organic phosphate compounds as well as of inorganic phosphate, which is an important controlling factor in metabolism, can be measured. Moreover, ^{31}P NMR is the only means for studying phosphorylated compounds in vivo without destroying the cells, thereby also giving information about intracellular compartmentation and pH values.[90,91] Few reports, however, describe the application of ^{31}P NMR to higher plants, and to our knowledge only one study existed for green algae, chiefly concerning polyphosphates in intact cells of *Cosmarium*.[92]* Therefore, the aim of our first studies had to be to test the

* In the meantime, ^{31}P NMR studies have also been performed with *Chlorella*, dealing with pH variation in chloroplast and cytoplasm of photosynthesizing cells,[93,94] with energy metabolism under aerobic and anaerobic conditions[106] or with polyphosphate distribution.[107]

Table 5
EFFECT OF METABOLITES ON
THE MAXIMAL VELOCITY
(V_{max}) OF PYRUVATE KINASE
ISOENZYMES OF YELLOW
***CHLORELLA* MUTANT NO. 20**

Addition	PK_1	PK_2
None	1.00	1.00
AMP	1.00	1.09
ATP	0.40	0.71
P_i	0.96	1.19
Citrate	0.42	0.62
F-6-P	0.91	0.96
FbP	0.95	1.10

Note: Final concentration of metabolites is 12 mM. V_{max} of the control is set 1.

After Ruyters, G., *Z. Pflanzenphysiol.*, 108, 207, 1982. With permission.

Table 6
EFFECT OF METABOLITES ON
PEP AFFINITY ($S_{0.5}$) OF
PYRUVATE KINASE
ISOENZYMES OF YELLOW
***CHLORELLA* MUTANT NO. 20**

Addition	PK_1	PK_2
None	1.00	1.00
AMP	0.55	0.44
ATP	1.82	3.23
P_i	1.23	1.65
Citrate	1.57	5.47
F-6-P	1.12	1.04
FbP	1.18	1.43

Note: Final concentration is 12 mM. $S_{0.5}$ of the control is set 1.

After Ruyters, G., *Z. Pflanzenphysiol.*, 108, 207, 1982. With permission.

adaptability of the ^{31}P NMR method to the study of extracts and whole cells of green algae with special emphasis on the detection, identification, and quantitative determination of phosphate intermediates.

Scenedesmus mutant C-2A′, widely used for the study of BL effects on carbon metabolism and chloroplast development,[47,58,95,96] was also chosen for the ^{31}P NMR spectroscopy. In ^{31}P NMR spectra of intact cells and extracts, a number of peaks were detected and identified as signals from sugar phosphates including G-6-P and FbP, inorganic phosphate, nucleotide di- and triphosphates, NAD(P), and UDPG. In the case of intact cells, additionally poly-phosphates were found.[88,89,105]

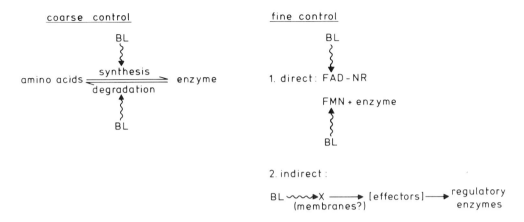

FIGURE 3. Control mechanisms of enzyme capacity and enzyme activity exerted by blue light (NR = nitrate reductase). (Modified from Ruyters, G., in *Blue Light Effects in Biological Systems*, Senger, H., Ed., Springer-Verlag, Berlin, 1984.)

To gain deeper insight into the regulation of carbohydrate breakdown in darkness and BL, ^{31}P NMR studies are in progress with special emphasis on the time course of the changes. First results indicate a transient drop in P_i, a quick decrease in the ATP/ADP ratio, and an increase in FbP and G-6-P after the onset of the illumination or after adding exogenous glucose, pointing to an important regulatory role of hexokinase and PFK.[88,105] (Since pyruvate, the product of the PK reaction, is not a phosphorylated compound, its possible changes cannot be measured with ^{31}P NMR; the level of its substrate PEP is usually very low and therefore cannot always be detected with ^{31}P NMR. Thus, no new information about the regulatory role of PK can be obtained by this method at the moment.)

After successfully introducing ^{31}P NMR spectroscopy into the investigation of phosphate intermediates of *Scenedesmus* C-2A′, we will continue these studies, especially in order to obtain more information about very quick changes in the level of metabolites.

Considering these results, it seems that BL can in fact exert fine control on different levels; however, the picture is far from being complete. Even for a proper understanding of the regulation of carbohydrate degradation, probably the best-studied pathway in BL research, much more information is needed. For a complete insight into the BL-mediated regulation of metabolism, the level of all intermediates and effectors and a time course for their changes have to be measured; so much remains to be done in elucidating the fine control of metabolism by blue light.

IV. CONCLUSION

Figure 3 tries to summarize the discussed results. Enzymes are certainly influenced by BL, although it has been proved only in a few cases by means of action spectroscopy that the response obtained is solely due to the operation of the BL photoreceptor(s) and not to that of phytochrome or chlorophyll. In most studies the maximal velocity V_{max} of an enzymatic reaction has been measured under optimal in vitro conditions, revealing the amount of active enzyme, the enzyme capacity. BL-mediated changes in V_{max}, together with protein synthesis inhibitor studies, strongly suggest an effect of BL on protein synthesis. For a few proteins this conclusion is supported by the demonstrated enhancement of the respective mRNA species; hopefully, more such reports will appear which prove the ability of the short-wave visible radiation to exert coarse control of metabolic pathways via changes in enzyme amount. Coarse control of metabolism is characteristic for development processes; in this context BL-regulated chloroplast development has to be considered above all.

BL also is able to influence the activity of already existing enzymes and thus to exert fine control. The enzyme molecule can be affected directly via its prosthetic group as in the case of nitrate reductase, or made sensitive to BL by the addition of FMN. Probably more important from the physiological point of view is the BL-mediated regulation of enzyme activity via changes in the level of substrates, products, coenzymes (including ions) or in the concentration of effectors, thereby quickly modulating regulatory enzymes and thus flux through metabolic pathways. However, not much information is available on this latter mechanism. Only the regulation of carbohydrate degradation of the yellow *Chlorella* mutant no. 20 has been studied in some detail with special emphasis on the regulatory enzyme PK. A new approach is currently being made by the use of ^{31}P NMR spectroscopy for measuring important intermediates. Since it is commonly accepted that the shorter is the time between light absorption and an observed response the more likely is a relatively direct relationship between photoreceptor and response, especially fast changes in metabolism should be studied intensively. Therefore, one main task for future BL research should be to obtain more information about fast changes in the level of substrates, products, coenzymes, and regulatory effectors, with special regard also to the compartmentation of these compounds, as well as about changes in pH and the content of important ions, which also play a role in the regulation of metabolism. Deeper insight into the primary mode of action of short-wavelength light on metabolism could thus be gained.

ACKNOWLEDGMENTS

The author wants to thank Prof. Dr. W. Kowallik for providing unpublished data and for valuable discussions and Mr. R. Church for reading the English. Part of his own work presented was supported by grants from the Japan Society for the Promotion of Science, the Alexander von Humboldt Foundation, and from the University of Bielefeld (special project nos. 2059 and 2079).

REFERENCES

1. **Senger, H.,** *The Blue Light Syndrome,* Springer-Verlag, Berlin, 1980.
2. **Senger, H.,** *Blue Light Effects in Biological Systems,* Springer-Verlag, Berlin, 1984.
3. **Senger, H. and Briggs, W. R.,** The blue light receptor(s): primary reactions and subsequent metabolic changes, *Photochem. Photobiol. Rev.,* 6, 1, 1981.
4. **Kowallik, W.,** Blue light effects on respiration, *Annu. Rev. Plant Physiol.,* 33, 51, 1982.
5. **Kowallik, W.,** Blue light effects on carbohydrate and protein metabolism, in *Blue Light Responses: Phenomena and Occurrence in Plants and Microorganisms,* Vol. 1., Senger, H., Ed., CRC Press, Boca Raton, Fla., 1987, chap. 2.
6. **Miyachi, S., Miyachi, S., and Kamiya, A.,** Wavelength effects on photosynthetic carbon metabolism in *Chlorella, Plant Cell Physiol.,* 19, 277, 1978.
7. **Voskresenskaya, N. P.,** Effect of light quality on carbon metabolism, in *Encyclopedia of Plant Physiology,* New Series, Vol. 6, Gibbs, M. and Latzko, E., Eds., Springer-Verlag, Berlin, 1979, 174.
8. **Conradt, W.,** Über die Wirkung kurzwellig-sichtbarer Strahlung auf plastidäre und cytoplasmatische Enzyme einer chlorophyllfreien *Chlorella* mutante, Ph.D. thesis, Universität Bielefeld, Bielefeld, W. Germany, 1980.
9. **Senger, H., Klein, O., Dörnemann, D., and Porra, R. J.,** The action of blue light on 5-aminolevulinic acid formation, in *The Blue Light Syndrome,* Senger, H., Ed., Springer-Verlag, Berlin, 1980, 541.
10. **Cerdá-Olmedo, E.,** Genetic determination of the responses of *Phycomyces* to light, in *Blue Light Effects in Biological Systems,* Senger, H., Ed., Springer-Verlag, Berlin, 1984, 220.

11. **Schmidt, G. and Lyman, H.,** Photocontrol of chloroplast enzyme synthesis in mutant and wild-type *Euglena gracilis,* in *Proc. 3rd Int. Congr. Photosynthesis,* Avron, M., Ed., Elsevier, Amsterdam, 1974, 1755.

12. **Kritsky, M. S., Afanasieva, T. P., Belozerskaya, T. A., Chailakhian, L. M., Chernysheva, E. K., Filippovich, S. Y., Levina, N. N., Potapova, T. V., and Sokolovsky, V. Y.,** Photoreceptor mechanism of *Neurospora crassa:* Control over the electrophysiological properties of cell membrane and over the level of nucleotide regulators, in *Blue Light Effects in Biological Systems,* Senger, H., Ed., Springer-Verlag, Berlin, 1984, 207.

13. **Cohen, R. J. and Atkinson, M. M.,** Activation of *Phycomyces* adenosine 3',5'-monophosphate phosphodiesterase by blue light, *Biochem. Biophys. Res. Commun.,* 83, 616, 1978.

14. **Feierabend, J.,** Development studies on microbodies in wheat leaves. III. On the photocontrol of microbody development, *Planta,* 123, 63, 1975.

15. **Horrum, M. A. and Schwartzbach, S. D.,** Nutritional regulation of organelle biogenesis in *Euglena:* photo- and metabolite induction of mitochondria, *Planta,* 149, 376, 1980.

16. **van der Velde, H. H., Guiking, P., and van der Wulp, D.,** Glucose-6-phosphate dehydrogenase and 6-phosphogluconate dehydrogenase in *Acrochaetium daviesii* cultured under red, white and blue light, *Z. Pflanzenphysiol.,* 76, 95, 1975.

17. **Eichhorn, M. and Augsten, H.,** Die Wirkung von Blau- und Rotlicht auf die Aktivität der Glucose-6-phosphate-Dehydrogenase und das Adenylatsystem bei *Wolffia arrhiza* unter steady state Bedingungen, *Z. Pflanzenphysiol.,* 85, 147, 1977.

18. **Kowallik, W. and Neuert, G.,** Enhancement by blue light of GOGAT activity in *Chlorella,* in *Blue Light Effects in Biological Systems,* Senger, H., Ed., Springer-Verlag, Berlin, 1984, 310.

19. **Stabenau, H.,** Aktivitätsänderungen von Enzymen bei *Chlorogonium elongatum* unter dem Einfluß von rotem und blauem Licht, *Z. Pflanzenphysiol.,* 67, 105, 1972.

20. **Kulandaivelu, G. and Sarojini, G.,** Blue light-induced enhancement in activity of certain enzymes in heterotrophically grown cultures of *Scenedesmus obliquus,* in *The Blue Light Syndrome,* Senger, H., Ed., Springer-Verlag, Berlin, 1980, 372.

21. **Steup, M.,** Über Beziehungen zwischen Kohlenhydrat-, Protein-, und Phosphatstoffwechsel bei *Chlamydomonas reinhardii* Dangeard im blauen und roten Spektralbereich, Ph.D. thesis, Universität Göttingen, Göttingen, West Germany, 1972.

22. **Conradt, W. and Ruyters, G.,** Blue light-effects on enzymes of the carbohydrate metabolism in *Chlorella.* II. Glyceraldehyde 3-phosphate dehydrogenase (NADP-dependent), in *The Blue Light Syndrome,* Senger, H., Ed., Springer-Verlag, Berlin, 1980, 368.

23. **Kowallik, W. and Grotjohann, N.,** Influence of blue light on NADP-dependent glyceraldehyde 3-phosphate dehydrogenase activity in *Chlorella,* in *Blue Light Effects in Biological Systems,* Senger, H., Ed., Springer-Verlag, Berlin, 1984, 302.

24. **Poyarkova, N. M., Drozdova, I. S., and Voskresenskaya, N. P.,** Effects of blue light on the activity of carboxylating enzymes and NADP$^+$-dependent glyceraldehyde 3-phosphate dehydrogenase in bean and maize plants, *Photosynthetica,* 7, 58, 1973.

25. **Gnanam, A., Habib Mohamed, A., and Seetha, R.,** Comparative studies on the effect of ammonia and blue light on the regulation of photosynthetic carbon metabolism in higher plants, in *The Blue Light Syndrome,* Senger, H., Ed., Springer-Verlag, Berlin, 1980, 435.

26. **Voskresenskaya, N. P., Grishina, G. S., Sechenska, M., and Drozdova, I. S.,** After-effect of blue and red light on oxidation of glycolic acid by pea chloroplasts and homogenates, *Sov. Plant Physiol.,* 17, 859, 1970.

27. **Voskresenskaya, N. P. and Khodzhiev, A. K.,** Aftereffect of red and blue light on activity of glycolate oxidase and glyoxylate aminotransferases in plants, *Sov. Plant Physiol.,* 20, 254, 1973.

28. **Tchang, F., Lecharny, A., and Mazliak, P.,** Photostimulation of hydroxypyruvate reductase activity in peroxisomes of *Pharbitis nil* seedlings. II. Photoreceptors in blue light, *Plant Cell Physiol.,* 25, 1039, 1984.

29. **Stein-Ludolph, G. and Kowallik, W.,** Inhibition by blue light of α-ketoglutarate dehydrogenase activity in a chlorophyll-free *Chlorella* mutant, *Biochem. Physiol. Pflanz.,* 180, 163, 1985.

30. **Afanasieva, T. P., Filippovich, S. Y., Sokolovsky, V. Y., and Kritsky, M. S.,** Developmental regulation of NAD$^+$ kinase in *Neurospora crassa, Arch. Microbiol.,* 133, 307, 1982.

31. **Richter, G., Hundrieser, J., Gross, M., Schultz, S., Bottländer, K., and Schneider, C.,** Blue light effects in cell cultures, in *Blue Light Effects in Biological Systems,* Senger, H., Ed., Springer-Verlag, Berlin, 1984, 387.

32. **Stevens, S. E. and van Baalen, C.,** Control of nitrate reductase in a blue-green alga, *Arch. Biochem. Biophys.,* 161, 146, 1974.

33. **Azuara, M. P. and Aparicio, P. J.,** In vivo blue-light activation of *Chlamydomonas reinhardii* nitrate reductase, *Plant Physiol.,* 71, 286, 1983.

34. **Jones, R. W. and Sheard, R. W.,** Effect of blue and red light on nitrate reductase level in leaves of maize and pea seedlings, *Plant Sci. Lett.,* 8, 305, 1977.

35. **Wild, A. and Holzapfel, A.,** The effect of blue and red light on the content of chlorophyll, cytochrome f, soluble reducing sugars, soluble proteins and the nitrate reductase activity during growth of the primary leaves of *Sinapis alba,* in *The Blue Light Syndrome,* Senger, H., Ed., Springer-Verlag, Berlin, 1980, 444.

36. **Stoy, V.,** Action of different light qualities on simultaneous photosynthesis and nitrate assimilation in wheat leaves, *Physiol. Plant.,* 8, 963, 1955.

37. **Rao, L. V. M., Datta, N., Guha-Mukherjee, S., and Sopory, S. K.,** The effect of blue light on the induction of nitrate reductase in etiolated excised maize leaves, *Plant Sci. Lett.,* 28, 39, 1982.

38. **Strotmann, H.,** Blaulichteffekte auf die Nitritreduktase von *Chlorella, Planta,* 73, 376, 1967.

39. **Attridge, T. H. and Smith, H.,** Density-labeling evidence for the blue-light-mediated activation of phenylalanine ammonium lyase in *Cucumis sativus* seedlings, *Biochim. Biophys. Acta,* 343, 452, 1974.

40. **Duke, S. O. and Naylor, A. W.,** Light control of anthocyanin biosynthesis in *Zea* seedlings, *Physiol. Plant.,* 37, 62, 1976.

41. **Ogasawara, N. and Miyachi, S.,** Regulation of CO_2 fixation in *Chlorella* by light of varied wavelengths and intensities, *Plant Cell Physiol.,* 11, 1, 1970.

42. **Ruyters, G.,** Blue light-enhanced phosphoenolpyruvate carboxylase activity in a chlorophyll-free *Chlorella* mutant, *Z. Pflanzenphysiol.,* 100, 107, 1980.

43. **Kamiya, A. and Miyachi, S.,** Blue light-induced formation of phosphoenolpyruvate carboxylase in colorless *Chlorella* mutant cells, *Plant Cell Physiol.,* 16, 729, 1975.

44. **Schmid, R.,** Blue light effects on morphogenesis and metabolism in *Acetabularia,* in *Blue Light Effects in Biological Systems,* Senger, H., Ed., Springer-Verlag, Berlin, 1984, 419.

45. **Kowallik, W. and Ruyters, G.,** Über Aktivitätssteigerungen der Pyruvat-Kinase durch Blaulicht oder Glucose bei einer chlorophyll-freien *Chlorella*- Mutante, *Planta,* 128, 11, 1976.

46. **Ruyters, G. and Miyachi, S.,** Changes in plastidic and cytoplasmic pyruvate kinase activity during chloroplast development of pea in blue and red light, *Plant Cell Physiol.,* 24, 863, 1983.

47. **Ruyters, G.,** Effects of blue light on pyruvate kinase activity during chloroplast development of unicellular green algae, *Photochem. Photobiol.,* 35, 229, 1982.

48. **Codd, G. A. and Stewart, R.,** The photoinactivation of microalgal ribulose bisphosphate carboxylase, in *The Blue Light Syndrome,* Senger, H., Ed., Springer-Verlag, Berlin, 1980, 392.

49. **Voskresenskaya, N. P.,** Control of the activity of photosynthetic apparatus in higher plants, in *Blue Light Effects in Biological Systems,* Senger, H., Ed., Springer-Verlag, Berlin, 1984, 407.

50. **Hundrieser, J. and Richter, G.,** Blue light-induced synthesis of ribulosebisphosphate carboxylase in cultured plant cells, *Plant Cell Rep.,* 1, 115, 1982.

51. **Vettermann, W.,** Mechanism of the light-dependent accumulation of starch in chloroplasts of *Acetabularia,* and its regulation, *Protoplasma,* 76, 261, 1973.

52. **Harding, R. W. and Shropshire, W.,** Photocontrol of carotenoid biosynthesis, *Annu. Rev. Plant Physiol.,* 31, 217, 1980.

53. **Rau, W.,** Blue light-induced carotenoid biosynthesis in microorganisms, in *The Blue Light Syndrome,* Senger, H., Ed., Springer-Verlag, Berlin, 1980, 283.

54. **Rau, W. and Schrott, E. L.,** Light-mediated biosynthesis in plants, *Photochem. Photobiol.,* 30, 755, 1979.

55. **Schrott, E. L.,** Carotenogenesis, in *Blue Light Effects in Biological Systems,* Senger, H., Ed., Springer-Verlag, Berlin, 1984, 366.

56. **Ruyters, G. and Kowallik, W.,** Further studies of the light-mediated change in the activity of pyruvate kinase of a chlorophyll-free *Chlorella* mutant, *Z. Pflanzenphysiol.,* 96, 29, 1980.

57. **Ruyters, G.,** Effects of blue light on enzymes, in *Blue Light Effects in Biological Systems,* Senger, H., Ed., Springer-Berlin, 1984, 283.

58. **Brinkmann, G. and Senger, H.,** Blue light regulation of chloroplast development in *Scenedesmus* mutant C-2A′, in *The Blue Light Syndrome,* Senger, H., Ed., Srpinger-Verlag, Berlin, 1980, 526.

59. **Hase, E.,** Effects of blue light on greening in microalgae, in *The Blue Light Syndrome,* Senger, H., Ed., Springer-Verlag, Berlin, 1980, 512.

60. **Ruyters, G.,** Effects of blue light on respiration and enzyme activity in a yellow *Chlorella* mutant, in *Proc. 5th Int. Congr. Photosynthesis,* Vol. 5, Akoyunoglou, G., Ed., Balaban International Science Services, Philadelphia, 1981, 905.

61. **Schiff, J. A. and Schwartzbach, S. D.,** Photocontrol of chloroplast development in *Euglena,* in *The Biology of Euglena,* Vol. 3, Buetow, D., Ed., Academic Press, New York, 1982, 313.

62. **Queiroz, O.,** Rhythms of enzyme capacity and activity as adaptive mechanisms, in *Encyclopedia of Plant Physiology,* New Series, Vol. 6, Gibbs, M. and Latzko, E., Eds., Springer-Verlag, Berlin, 1979, 126.

63. **Ap Rees, T.,** Integration of pathways of synthesis and degradation of hexose phosphates, in *The Biochemistry of Plants,* Vol. 3, Preiss, J., Ed., Academic Press, New York, 1980, 1.

64. **Rolleston, F. S.,** A theoretical background to the use of measured concentrations of intermediates in study of the control of intermediary metabolism, *Curr. Top. Cell. Regul.,* 5, 47, 1972.

65. **Newsholme, E. A. and Start, C.,** *Regulation in Metabolism,* John Wiley & Sons, New York, 1973.

66. **Heinrich, R., Rapoport, S. M., and Rapoport, T. A.,** Metabolic regulation and mathematical models, *Prog. Biophys. Mol. Biol.,* 32, 1, 1977.

67. **Richter, G.,** Blue light effects on the level of translation and transcription, in *Blue Light Effects in Biological Systems,* Senger, H., Ed., Springer-Verlag, Berlin, 1984, 253.

68. **Ninnemann, H.,** The nitrate reductase system, in *Blue Light Effects in Biological Systems,* Senger, H., Ed., Springer-Verlag, Berlin, 1984, 95.

69. **Aparicio, P. J. and Maldonado, J. M.,** Photochromic regulation of nitrate reductase in green algae and higher plants, in *Molecular Models of Photoresponsiveness,* Montagnoli, G. and Erlanger, B. G., Eds., Plenum Press, New York, 1983, 364.

70. **Schmid, G. H.,** The effect of blue light on some flavin enzymes, *Hoppe-Seyler's Z. Physiol. Chem.,* 351, 575, 1970.

71. **Schmid, G. H.,** Photoregulation of β-D-glucose oxidase by blue light, *Phytochemistry,* 10, 2041, 1971.

72. **Schmid, G. H., and Schwarze, P.,** Blue light enhanced respiration in a colorless *Chlorella* mutant, *Hoppe Seyler's Z. Physiol. Chem.,* 350, 1513, 1969.

73. **Schmid, G. H.,** The effect of blue light on glycolate oxidase of tobacco, *Hoppe-Seyler's Z. Physiol. Chem.,* 350, 1035, 1969.

74. **Codd, G. A.,** The photoinhibition of malate dehydrogenase, *FEBS Lett.,* 20, 211, 1972.

75. **Vargas, M. A., Maurino, S. G., Maldonado, J. M., and Aparicio, P. J.,** Photoinactivation of spinach nitrate reductase sensitized by flavin mononucleotide. Evidence for the involvement of singlet oxygen, *Photochem. Photobiol.,* 36, 223, 1982.

76. **Codd, G. A.,** The photoinactivation of tobacco transketolase in the presence of flavin mononucleotide, *Z. Naturforsch.,* 27b, 701, 1972.

77. **Angele, S.,** Über die Wirkung von Blaulicht auf Kohlenhydrat abbauende Enzyme bei *Chlorella* , Ph.D. thesis, Universität Bielefeld, Bielefeld, W. Germany, 1981.

78. **Kowallik, W. and Gaffron, H.,** Respiration induced by blue light, *Planta,* 69, 92, 1966.

79. **Ruyters, G.,** Über Aktivitätsänderungen Kohlenhydrat-abbauender Enzyme von *Chlorella* im Dunkel und im Blaulicht, Ph.D. thesis, Universität Köln, Cologne, W. Germany, 1977.

80. **Kirsch, M.,** Nachweis von Enzymen des Tricarbonsäurecyclus bei einer chlorophyllfreien *Chlorella*- Mutante und ihre Beeinflussung durch Aussenfaktoren, Ph.D. thesis, Universität Bielefeld, Bielefeld, W. Germany, 1981.

81. **Kowallik, W. and Gaffron, H.,** Enhancement of respiration and fermentation in algae by blue light, *Nature,* 215, 1038, 1967.

82. **Turner, J. F. and Turner, D. H.,** The regulation of glycolysis and pentose phosphate pathway, in *Biochemistry of Plants,* Vol. 2, Davies, D. D., Ed., Academic Press, New York, 1980, 279.

83. **Kowallik, W. and Scheil, I.,** Lichtbedingte Veränderungen des ATP-Spiegels einer chlorophyllfreien *Chlorella*- Mutante, *Planta,* 131, 105, 1976.

84. **Ruyters, G.,** Isoenzymes of pyruvate kinase in a chlorophyll-free *Chlorella* mutant and their blue-light-mediated interconversion, *Z. Pflanzenphysiol.,* 103, 109, 1981.

85. **Ruyters, G.,** Regulatory properties of pyruvate kinase isoenzymes from a chlorophyll-free *Chlorella* mutant in darkness or blue-light, *Z. Pflanzenphysiol.,* 108, 207, 1982.

86. **Engström, L.,** Regulation of liver pyruvate kinase by phosphorylation-dephosphorylation, in *Recently Discovered Systems of Enzyme Regulation by Reversible Phosphorylation,* Cohen, P., Ed., Elsevier, Amsterdam, 1980, 31.

87. **Steup, M. and Pirson, A.,** Über den Einfluss des blauen und roten Spektralbereichs auf Phosphatfraktionen, besonders Polyphosphate, bei Grünalgen, *Biochem. Physiol. Pflanz.,* 166, 447, 1974.

88. **Ruyters, G., Oh-hama, T., and Kowallik, W.,** Phosphate compounds of *Scenedesmus* C-2A' in darkness or blue light as measured by ^{31}P NMR, *Plant Cell Physiol.,* 26, 571, 1985.

89. **Oh-hama, T., Ruyters, G., Furihata, K., Seto, H., and Miyachi, S.,** ^{31}P NMR studies in *Scenedesmus* C-2A' in darkness and blue light, in *Blue Light Effects in Biological Systems,* Senger, H., Ed., Springer-Verlag, Berlin, 1984, 323.

90. **Gadian, D. G., Radda, G. K., Richards, R. E., and Seely, P. J.,** ^{31}P NMR in living tissue: the road from a promising to an important tool in biology, in *Biological Application of Magnetic resonance,* Shulman, R. G., Ed., Academic Press, New York, 1979, 463.

91. **Roberts, J. K. M.,** Study of plant metabolism in vivo using NMR spectroscopy, *Annu. Rev. Plant Physiol.,* 35, 375, 1984.

92. **Elgavish, A., Elgavish, G. A., and Halmann, M.,** Intracellular phosphorous pools in intact algal cells, *FEBS Lett.,* 117, 137, 1980.

93. **Sianoudis, J., Mayer, A., and Leibfritz, D.,** Investigation of intracellular phosphate pools of the green alga *Chlorella* using ^{31}P nuclear magnetic resonance, *Org. Magn. Reson.,* 22, 364, 1984.

94. **Mitsumori, F. and Ito, O.,** Phosphorus-31 nuclear magnetic resonance studies of photosynthesizing *Chlorella, FEBS Lett.,* 174, 248, 1984.
95. **Oh-hama, T. and Senger, H.,** The development of structure and function in chloroplasts of greening mutants of *Scenedesmus.* III. Biosynthesis of δ-aminolevulinic acid, *Plant Cell Physiol.,* 16, 395, 1975.
96. **Watanabe, M., Oh-hama, T., and Miyachi, S.,** Light induced carbon metabolism in an early stage of greening in wild type and mutant C-2A′ cells of *Scenedesmus obliquus,* in *The Blue Light Syndrome,* Senger, H., Ed., Springer-Verlag, Berlin, 1980, 332.
97. **Jan, Y. N.,** Properties and cellular localization of chitin synthetase in *Phycomyces blakesleeanus, J. Biol. Chem.,* 249, 1973, 1974.
98. **Steinmüller, K. and Zetsche, K.,** Photo- and metabolite regulation of the synthesis of ribulose bisphosphate carboxylase/oxygenase and the phycobiliproteins in the alga *Cyanidium caldarium, Plant Physiol.,* 76, 935, 1984.
99. **Rau, W.,** Photoregulation of carotenoid biosynthesis, in *Biosynthesis of Isoprenoid Compounds,* Porter, J. W. and Spurgeon, S. L., Eds,. John Wiley & Sons, New York, 1983, 123.
100. **Wellburn, A. R.,** Ultrastructural, respiratory and metabolic changes associated with chloroplast development, in *Chloroplast Biogenesis,* Baker, N. R. and Barber, J., Eds., *Topics in Photosynthesis,* Vol. 5, Elsevier, New York, 1984, 253.
101. **Richter, G.,** Blue light control of the level of two plastid mRNAs in cultured plant cells, *Plant Mol. Biol.,* 3, 271, 1984.
102. **Duke, S. H. and Duke, S. O.,** Light control of extractable nitrate reductase activity in higher plants, *Physiol. Plant.,* 62, 485, 1984.
103. **Kowallik, W. and Stein, G.,** personal communication, 1981.
104. **Kowallik, W. and Neuert, G.,** personal communication, 1983.
105. **Ruyters, G.,** Regulation of blue light-enhanced carbohydrate breakdown during chloroplast development of Scenedesmus mutant C-2A′: A ³¹P-NMR study, in *Regulation of Chloroplast Differentiation,* Akoyunoglou, G. and Senger, H., Eds., Alan R. Liss, New York, 1986, 677.
106. **Sianoudis, J., Küsel, A. C., Naujokat, T., Offermann, W., Mayer, A., Grimme, L. H., and Leibfritz, D.,** Respirational activity of *Chlorella fusca* monitored by in vivo P-31 NMR, *Eur. Biophys. J.,* 13, 89, 1985.
107. **Sianoudis, J., Küsel, A. C., Mayer, A., Grimme, L. H., and Leibfritz, D.,** Distribution of polyphosphates in cell-compartments of *Chlorella fusca* as measured by ³¹P-NMR-spectroscopy, *Arch. Microbiol.,* 144, 48, 1986.

Chapter 6

MEMBRANE-BOUND BLUE LIGHT RECEPTORS — POSSIBLE CONNECTION TO BLUE LIGHT PHOTOMORPHOGENESIS

S. Widell

TABLE OF CONTENTS

I. Introduction 90

II. Physiological Relevance of Light-Induced Absorbance Changes 90

III. Membrane Purification 91

IV. Is There More Than One LIAC? 91

V. Inhibitor Studies 94

VI. Diversity of LIACs 95

VII. Final Remarks 96

Acknowledgments 96

References 96

I. INTRODUCTION

Light provides plants not only with energy for photosynthesis but also with information that is used in their development. Phytochrome, the "blue-light receptor", and UV receptors are some of the pigments acting as photoreceptors involved in developmental processes.

Phototropism, phototaxis, and chloroplast orientation are examples of blue-light-mediated processes. The action spectra for these processes are very similar, with broad band in the blue (with the main peak at 450 to 460 nm) and another in the UV region (around 370 nm). The similarity among action spectra has led to a search for a ubiquitous blue light receptor analogous to phytochrome, the red light receptor. Carotenoids and flavins have been suggested, but the problem as stated has not been solved yet. Rather, the suggestion that there should be only one photoreceptor in the blue region needs to be reevaluated.

The processes mentioned above are all characterized by being a change in orientation. This implies that the photoreceptor, at least when active, is bound to a membrane fraction in order to exert its effect. Therefore, there has since long been a search for blue light-sensitive pigments in the various membrane fractions of the cell.

The search for potential photoreceptors began with the use of highly sensitive dual-wavelength spectrophotometers such as those used in the detection of phytochrome.[1] The first success occurred in 1974 when blue-light-induced absorbance changes (LIAC) could be detected in *Dictyostelium discoideum* and *Phycomyces blakesleeanus*.[2] Blue light, with maximal activity around 460 nm and a second peak of activity at 360 nm, was found to induce absorbance changes, reflecting the reduction of a b-type cytochrome.[3,4] Furthermore, these changes appeared to be reversible; i.e., the cytochrome was reoxidized in the darkness. A flavin (or flavoprotein) was suggested to be the photoreceptor since excited flavins are able to photoreduce cytochromes in artificial systems.[5] Flavins undergo and mediate redox changes much more readily than do carotenoids (which rather isomerize as a function of light[6]).

A similar b-type cytochrome was also found in other fungi (e.g., *Neurospora crassa*[3]) as well as in higher plants.[7] In fungi, it was located in a pelletable membrane fraction, which was neither enriched in mitochondria nor in endoplasmic reticulum. In higher plants, the cytochrome showed only transient light sensitivity, however, and was therefore difficult to quantify.[7] It was thought to be identical to a dithionite- but not NADH-reducible b cytochrome located in a similar membrane fraction as that for fungi.[8]

II. PHYSIOLOGICAL RELEVANCE OF LIGHT-INDUCED ABSORBANCE CHANGES

Much doubt has been raised about the relationship between LIAC and blue light physiology, since flavins very easily undergo photoreduction and flavin-cytochrome complexes are common to almost all membranes of the cell. Furthermore, flavins could be released during cell disruption and membrane fractionation and then become more susceptible to light. LIAC has also been evoked in many organisms, even in species without known photosensitivity; e.g., HeLa cells.[9] Furthermore, the quantum yield of blue light-mediated cytochrome b reduction was found to be as low as 0.015 in *Phycomyces* mutants.[9]

Support for a connection between LIAC and physiology comes from experiments with artificial photoreceptors. In the presence of methylene blue, carotenoid synthesis in *Fusarium aqueductuum* (a normally blue light-regulated process) could be induced by red light.[10] This approach was used with corn coleoptiles, where methylene blue could substitute for the endogenous photoreceptor in mediating LIAC,[11] although the photochemistry of flavins and methylene blue is not the same.[12] The reaction was localized in a pelletable fraction that was neither enriched in mitochondria nor in endoplasmic reticulum,[11] similar to the one described above[7,8] that was enriched in light-reducible cytochrome b. Furthermore, methylene blue could substitute for the endogenous receptor both in mediating LIAC and in mediating phase shifts in conidiation in *Neurospora*.[13] It was later found that stable LIAC could be obtained also with blue light in higher plants if only the oxygen tension was lowered; e.g., by including glucose and glucose oxidase in the assay medium.[14] This LIAC had a similar distribution between membrane fractions as the methylene-blue-mediated LIAC.

III. MEMBRANE PURIFICATION

Another approach to elucidating the physiological role of LIAC has been to purify the membrane fraction to which it belongs. This work has proceeded at two different laboratories with totally different methods. Sucrose gradient centrifugation followed by Renografin gradient centrifugation resulted in a fraction rich in LIAC that contained hardly any other cytochromes as judged by low-temperature spectroscopy.[15,16] This fraction was identified with the plasma membrane since it showed glucan synthetase II activity as well as Mg^{2+}-ATPase activity in the presence of oligomycin.[15]

It has long been a problem to identify the plasma membrane as purified by gradient centrifugation. Several markers have been proposed, such as K^+, Mg^{2+}-ATPase, glucan synthetase II, and naphthyl phthalamic acid binding and staining with silicotungstic acid, but none of them have been applicable to all plants tested.[17] With aqueous polymer two-phase partition, the separation of membranes is based on differences in surface properties of the membranes rather than differences in size and density (as in gradient centrifugation). The procedure for phase partition is as follows: membranes are added to a system with two different aqueous phases (obtained by mixing aqueous solutions of dextran and polyethylene glycol, salt, sucrose, and buffer in desired concentrations). The upper, polyethylene glycol-enriched phase will have properties that are different from the lower, dextran-enriched phase, and membrane vesicles will partition between the two phases according to the properties of the surface exposed to the phase system. By varying polymer and ion concentrations a phase system optimal for membrane separation can be obtained. Phase partition has earlier been successfully used for the purification of chloroplasts, mitochondria, etc.[18] It has also been used in the purification of protoplasts,[19] i.e., particles which expose the plasma membrane to the medium, and which were found to have a very high affinity to the *upper* phase in these phase systems. In contrast, intracellular membranes have a low partition.

Phase partition could provide an additional identification method for plasma membranes since the partition behavior for protoplasts is known.[19] Plasma membranes from corn, barley, wheat, and cauliflower were successfully separated from mitochondria and other intracellular membranes.[20,21] These plasma membranes had a similarly high affinity for the upper phase as protoplasts. LIAC and plasma membrane markers, such as glucan synthetase II and K^+, Mg^{2+}-ATPase, were also collected here.[22] Furthermore, all the vesicles in the plasma membrane fraction were well stained with silicotungstic acid, similar to the staining of the plasma membranes in whole tissue section micrographs.[22] All the plasma membrane vesicles recovered from the upper phase were right-side-out and closed, as judged by the latencies of K^+, Mg^{2+}-ATPase and glucan synthetase II[23] and as might be expected from their similarity in partition behavior with intact protoplasts. Thus, LIAC is located in the plasma membranes. The action spectrum of this LIAC (Figure 1) was similar to that for typical blue light responses such as phototropism and also similar to absorption spectra of flavins.[24]

IV. IS THERE MORE THAN ONE LIAC?

A fraction rich in intracellular membranes (a fraction having a high affinity for the lower phase) contains LIAC as well as a part of the putative plasma membrane markers.[20,22] This

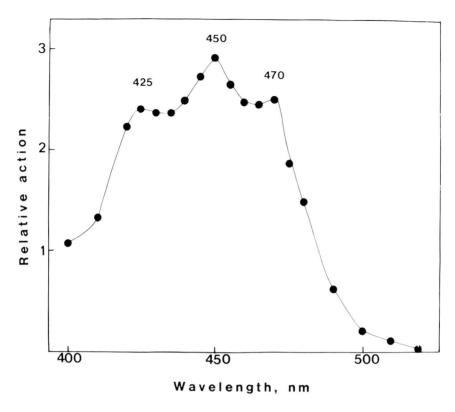

FIGURE 1.   Action spectrum of LIAC (blue light-induced absorbance change, $\Delta(A_{430}\text{-}A_{410})$ in a plasma membrane fraction from cauliflower. Protein concentration: 0.4 mg/mℓ.[24]

could imply that LIAC resides in more than one membrane fraction, which might indicate that the reaction is not as specific as suggested.[14,15] It could also be that plasma membranes, when broken into vesicles, turn either their original right-side-out or become everted. The LIAC found in the lower phase could then be ascribed to everted plasma membrane vesicles (if such are obtained during homogenization), since the plasma membrane fraction recovered from the upper phase exclusively contained right-side-out-vesicles.[23] It also could not be ruled out that plasma membranes with very different surface properties exist in the tissues, which also would explain a dual distribution of plasma membrane markers in the phase system.

To determine whether the LIAC found in the plasma membrane is specific to that membrane, the lower-phase LIAC needs to be further analyzed. Membranes were prepared from three different centrifuge fractions of cauliflower.[25] It was found that lower-phase LIAC copurified with endoplasmic reticulum since the ratios of lower-phase LIAC to NADPH cytochrome c reductase were about the same in these centrifuge fractions (except for 9 KP heavy fraction, Table 1).

The plasma membrane fraction from cauliflower (i.e., the fraction with a high affinity for the *upper* phase) was analyzed with low-temperature spectroscopy.[24] Only one cytochrome was detectable irrespective of whether the reduction was performed by light or by dithionite (Figure 2). The α-band was located at 556 nm (see also Reference 16), which is very similar to cytochrome P-450/420. Carbon monoxide difference spectra, as well as pyridine binding spectra, typical for cytochrome P-450 were also obtained with the plasma membrane fraction.[26] Mild lithium dodecyl sulfate polyacrylamide gel electrophoresis of the plasma membranes showed only one heme-staining band with an apparent molecular weight of 93 kDa, in agreement with a dimer of cytochrome P-450/420.[26] Cytochrome P-450 has

Table 1
SPECIFIC ACTIVITIES OF LIAC, CYTOCHROME C
OXIDASE, AND NADPH-DEPENDENT CYTOCHROME C
REDUCTASE IN DIFFERENT MEMBRANE FRACTIONS
OBTAINED FROM CAULIFLOWER INFLORESCENCES

Fraction	LIAC $\Delta(\Delta A)\ 10^3$	CCO nmol min^{-1}	CCR nmol min^{-1}	LIAC (CCO)	LIAC (CCR)
9 KP, PM	6.8	12	52	0.57	0.13
21 KP, PM	13.2	6	44	2.2	0.30
50 KP, PM	15.6	9	26	1.7	0.60
9 KP, IM					
Light	3.0	830	44	0.004	0.068
Heavy	2.1	1840	113	0.001	0.019
21 KP, IM	5.0	650	60	0.008	0.083
50 KP, IM	3.6	330	66	0.011	0.055
50 KP	4.6	27	57	0.17	0.081

Note: Fractionation was achieved by differential centrifugation, giving 9 KP, (9,000-g pellet), 21 KP (21,000-g 20-min pellet), and 50 KP (50,000-g 45-min pellet) fractions, followed by partition in an aqueous polymer two-phase system, giving PM (plasma membranes) and IM (intracellular membranes). 9 KP, IM was further fractionated into a light and a heavy band (intact mitochondria) on a Percoll gradient.[25] CCO, cytochrome c oxidase; CCR, NADPH-cytochrome c reductase.

been well studied in animal and bacterial systems. It usually acts as a mixed-function oxidase and is part of an electron transport chain starting with NADPH and including flavoprotein (flavinadenin nucleotide [FAD] and flavin mononucleotide [FMN]). Cytochrome P-450 exists in many forms and may be induced by foreign toxic compounds. It has become more and more evident that this cytochrome also plays important metabolic roles in higher plants,[27,28] such as in hormone metabolism,[29] detoxification,[30] phenylpropanoid synthesis,[31] etc. Furthermore, NAD(P)H cytochrome c reductase was found associated with the plasma membrane.[21,25,32] In the cell, cytochrome P-450 is the physiological substrate for this enzyme. Thus, properties that earlier have been ascribed to the endoplasmic reticulum (and mitochondria in some cases), i.e., the presence of cytochrome P-450 and NADPH-cytochrome c reductase,[33] can also be ascribed to the plasma membrane. The membrane preparation obtained with phase partition clearly does not contain any endoplasmic reticulum, since cytochrome b_5 was not detected.[24]

The fraction rich in intracellular membranes (i.e., the fraction with a high affinity for the *lower* phase) was also analyzed with low-temperature spectroscopy. Chemical reduction revealed several cytochromes (Figure 3). At least three of these were light sensitive, with α-bands at 599, 557, and 551 nm, respectively. The band at 599 nm could be attributed to cytochrome c oxidase, whereas the two others are not readily identified. It is possible that they are identical to cytochrome b_5 from the endoplasmic reticulum, which has a split α-band.[27] It could not be ruled out that the α-bands come from two different cytochromes, cytochrome c_1 of mitochondria[34] and a b-cytochrome similar or identical to that in the plasma membrane fraction. If the latter is true, it indicates that there are two different populations of plasma membrane vesicles in the microsomal fraction, one going to the upper phase and another going to the lower phase upon phase partitioning.

In conclusion, there are several LIACs. One is located in the plasma membrane and has properties similar to cytchrome P-450. LIAC in the lower phase is probably located in the endoplasmic reticulum. However, it could not be ruled out that this LIAC, at least partially,

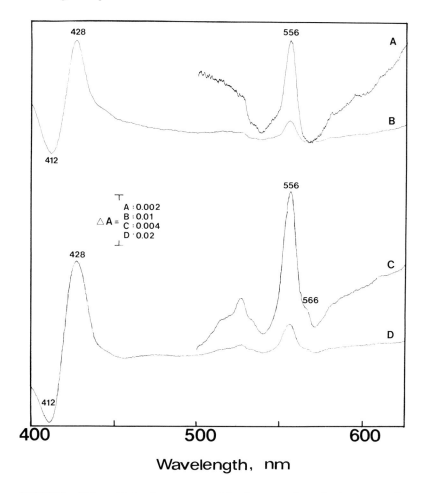

FIGURE 2.   Differential absorbance spectra of the plasma membrane fraction from cauliflower. A,B: absorbance of blue-irradiated sample minus absorbance of sample in dark. C,D: absorbance of dithionite-reduced sample minus absorbance of oxidized sample. Temperature 77 K. Protein concentration: 1 mg/mℓ.[24]

is related to plasma membranes that have other surface properties (and therefore behave differently in the phase system) compared to those recovered in the upper phase.

V. INHIBITOR STUDIES

More direct evidence for the involvement of the plasma-membrane-bound flavin-LIAC couple in the photobiology of higher plants comes from experiments with inhibitors. Potassium iodide, phenylacetic acid, and azide specifically inhibited phototropism compared to geotropism.[35] These inhibitors, which are known to interact with photoexcited flavins, also blocked the LIAC.[36,37]

The most elegant correlation was found with the diphenyl ether acifluorfen.[36] This substance stimulated phototropism in oats to a maximum of about 30%. A similar stimulation was found with LIAC (although not at the same acifluorfen concentration). The inhibitor interacted directly with LIAC by delaying the reoxidation of the cytochrome after irradiation. The maximum delay was found at a similar acifluorfen concentration as gave maximal stimulation of LIAC.

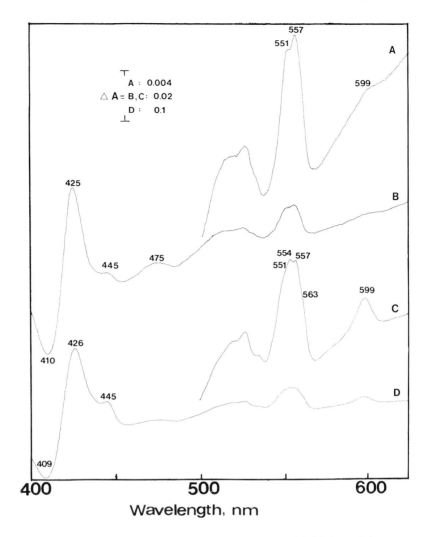

FIGURE 3. Differential absorbance spectra of the fraction enriched in intracellular membranes from cauliflower. A,B: absorbance of blue-irradiated sample minus absorbance of sample in dark. C,D: absorbance of dithionite-reduced sample minus absorbance of oxidized sample. Temperature 77 K. Protein concentration: 10 mg/mℓ.[24]

VI. DIVERSITY OF LIACs

Thus, flavin-cytochrome complexes could indeed be connected to blue light physiology. Whether there is more than one LIAC or not does not rule out this relation, since there are many different blue light responses. In *Neurospora,* blue light-stimulated conidiation seems to be regulated by nitrate reductase, another flavin cytochrome system that can show LIAC.[38,39] Starvation of mycelia leads to dissociation of the cytoplasmic nitrate reductase and association of the subunits with a fraction rich in plasma membranes.[39] At the same time, LIAC appears. The same fungus, although not the same strain, shows also photoinduced phase shifts, which are mediated by a plasma membrane-bound flavin-cytochrome complex that probably is different from nitrate reductase.[13,40] Another organism with more than one LIAC is *Dictyostelium discoideum.* At its multicellular stage, this mold shows in addition to the ''normal'' LIAC described above (light-induced reduction of a b-cytochrome, probably localized in the plasma membrane) a LIAC indicating the direct (not via flavin) photooxidation of a high-spin hemoprotein.[41] The difference spectrum obtained by irradiation of the mold shows a prominent peak at 430 nm and a broad shoulder at 550 to 590 nm which corresponds to the action spectrum for phototaxis. The pigment was purified by sonication of a mitochrondrial fraction. $(NH_4)_2SO_4$-precipitation of the soluble fraction followed by sucrose gradient centrifugation yielded a pigment enriched 2000-fold with a molecular weight of 240,000. The same organism, in its single-cell state, accumulates in light traps in a positively phototactic way.[42] The action spectrum is complex and could neither be matched solely to the high-spin hemoprotein nor to a flavoprotein.

There are many other photoprocesses that include more than one membrane and pigment; e.g., carotenoid synthesis of *Neurospora*,[43] and anthocyanin synthesis in higher plants.[44] Anthocyanin synthesis, a photoresponsive process which in some cases is blue light-mediated, involves cinnamic acid 4-hydroxylase, another cytochrome P-450-containing enzyme complex, which in this system is claimed to be bound to the endoplasmic reticulum.[45] Finally, photobiological research on different mutants of *Phycomyces blakesleeanus* has revealed that the sporangiophore must contain several closely arranged photoreceptors, interacting in a so far unknown manner in the phototropic reaction.[46]

VII. FINAL REMARKS

It is hoped that the identification of the plasma membrane-bound LIAC as a cytochrome P-450 will give some clues to the transduction of LIAC into a physiological response and perhaps also knowledge about the coordination between different pigments such as phytochrome and the blue light receptor, or UV-A and the blue light receptor. Of great importance in this discussion is the discovery that pteridines (pigments absorbing UV-A) associated with nitrate reductase seem to have photobiological functions.[47] Similar pteridine-like pigments are also found associated with plasma membranes of cauliflower and oat shoot,[37] and these could perhaps act as antennae to the flavoprotein.

ACKNOWLEDGMENTS

I wish to thank Professor T. Murphy, Department of Botany, Davis, California, for important comments on the manuscript as well as for linguistic help.

REFERENCES

1. **Butler, W. L., Norris, K. H., Siegelman, H. W., and Hendricks, S. B.,** Detection, assay and preliminary purification of the pigment controlling photoresponsive development of plants, *Proc. Natl. Acad. Sci. U.S.A.,* 45, 1703, 1959.
2. **Poff, K. and Butler, W. L.,** Absorbance changes induced by blue light in *Phycomyces blakesleeanus* and *Dictyostelium discoideum, Nature (London),* 248, 799, 1974.
3. **Muñoz, V., Brody, S., and Butler, W. L.,** Photoreceptor pigment for blue light responses in *Neurospora crassa, Biochem. Biophys. Res. Commun.,* 58, 322, 1974.
4. **Muñoz, V. and Butler, W. L.,** Photoreceptor pigment for blue light in *Neurospora crassa, Plant Physiol.,* 55, 421, 1975.
5. **Schmidt, W. and Butler, W. L.,** Flavin mediated photoreactions in artificial systems: a possible model for the blue light photoreceptor pigment in living systems, *Photochem. Photobiol.,* 24, 71, 1976.
6. **Steinitz, Y. L., Schiff, J. A., Osafune, T., and Green, M. S.,** *Cis* to *trans* photoisomerization of ζ-carotene in *Euglena gracilis* var. *bacillaris* W_3 BUL: further purification and characterization of the photoactivity, in *The Blue Light Syndrome,* Senger, H., Ed., Springer-Verlag, Berlin, 1980, 269.
7. **Brain, R. D., Freeberg, J. A., Weiss, C., and Briggs, W. R.,** Blue light induced absorbance changes in membrane fractions from corn and *Neurospora, Plant Physiol.,* 59, 948, 1977.

8. **Jesaitis, A. J., Heners, P. R., Hertel, R., and Briggs, W. R.,** Characterization of a membrane fraction containing a b-type cytochrome, *Plant Physiol.,* 59, 941, 1977.
9. **Lipson, E. D. and Presti, D.,** Light induced absorbance changes in *Phycomyces* photomutants, *Photochem. Photobiol.,* 25, 203, 1977.
10. **Lang-Feulner, J. and Rau, W.,** Redox dyes as artificial photoreceptors in light dependent carotenoid synthesis, *Photochem. Photobiol.,* 21, 179, 1975.
11. **Britz, S. J., Schrott, E., Widell, S., and Briggs, W. R.,** Red light induced reduction of a particle associated *b*-type cytochrome from corn in the presence of methylene blue, *Photochem. Photobiol.,* 29, 359, 1979.
12. **Widell, S., Britz, S. J., and Briggs, W. R.,** Characterization of the red light induced reduction of a particle associated *b*-type cytochrome from corn in the presence of methylene blue, *Photochem. Photobiol.,* 32, 669, 1980.
13. **Brain, R. D.,** as cited in **Senger, H. and Briggs, W. R.,** The blue light receptor(s): primary reactions and subsequent metabolic changes, in *Photochemical and Photobiological Reviews,* Vol. 6, Smith, K., Ed., Plenum Press, New York, 1980, 1.
14. **Goldsmith, M. H., Caubergs, R. J., and Briggs, W. R.,** Light inducible cytochrome reduction in membrane preparations from corn coleoptiles. I. Stabilization and spectral characterization of the reaction, *Plant Physiol.,* 66, 1067, 1980.
15. **Leong, T.-Y. and Briggs, W. R.,** Partial purification and characterization of a blue light-sensitive cytochrome-flavin complex from corn membranes, *Plant Physiol.,* 67, 1042, 1981.
16. **Leong, T.-Y., Vierstra, R. D., and Briggs, W. R.,** A blue light-sensitive cytochrome-flavin complex from corn coleoptiles. Further characterization, *Photochem. Photobiol.,* 34, 697, 1981.
17. **Quail, P. H.,** Plant cell fractionation, *Annu. Rev. Plant Physiol.,* 30, 425, 1979.
18. **Larsson, C.,** Partition in aqueous polymer two-phase systems — a rapid method for separation of membrane particles according to their surface properties, in *Isolation of Membranes and Organelles from Plant Cells,* Hall, J. L. and Moore, A. L., Eds., Academic Press, London, 1983, 277.
19. **Hallberg, M. and Larsson, C.,** Compartmentation and export of $^{14}CO_2$ fixation products in mesophyll protoplasts from the C_4 plant *Digitaria sanguinalis, Arch. Biochem. Biophys.,* 208, 121, 1981.
20. **Widell, S. and Larsson, C.,** Separation of presumptive plasma membranes from mitochondria by partition in an aqueous polymer two-phase system, *Physiol. Plant.,* 51, 368, 1981.
21. **Lundborg, T., Widell, S., and Larsson, C.,** Distribution of ATPases in wheat root membranes separated by phase partition, *Physiol. Plant.,* 52, 89, 1981.
22. **Widell, S. Lundborg, T., and Larsson, C.,** Plasma membrane from oats prepared by partition in an aqueous polymer two-phase system — on the use of light induced cytochrome *b* reduction as a marker for the plasma membrane, *Plant Physiol.,* 70, 1429, 1982.
23. **Larsson, C., Kjellbom, P., Widell, S., and Lundborg, T.,** Sidedness of plant plasma membranes vesicles purified by partitioning in aqueous two-phase systems, *FEBS Lett.,* 171, 271, 1984.
24. **Widell, S., Caubergs, R. J., and Larsson, C.,** Spectral characterization of light reducible cytochrome in a plasma membrane-enriched fraction and in other membranes from cauliflower influorescences, *Photochem. Photobiol.,* 38, 95, 1983.
25. **Widell, S. and Larsson, C.,** Distribution of cytochrome *b* photoreductions mediated by endogenous photosensitizer or methylene blue in fractions from corn and cauliflower, *Physiol. Plant.,* 57, 196, 1983.
26. **Kjellbom, P., Larsson, C., Askerlund, P., Schelin, C., and Widell, S.,** Cytochrome P-450/420 in plant plasma membranes: a possible component of the blue light reducible flavoprotein-cytochrome complex, *Photochem. Photobiol.,* 42, 779, 1985.
27. **Rich, P. R. and Bendall, D. S.,** Cytochrome components of plant microsomes, *Eur. J. Biochem.,* 55, 333, 1975.
28. **West, C. A.,** Hydroxylases, monooxygenases, and cytochrome P-450, in *The Biochemistry of Plants,* Vol. 2, Stumpf, P. K. and Conn. E. E., Eds., Academic Press, London, 1980, 317.
29. **Chen, C.-M. and Leisner, S. M.,** Modification of cytokinins by cauliflower microsomal enzymes, *Plant Physiol.,* 75, 442, 1984.
30. **Adele, P., Reichart, D., Salaun, J.-P., Benveniste, I., and Durst, F.,** Induction of cytochrome P-450 and monooxygenase activity by 2,4-dichlorophenoxyacid in higher plant tissue, *Plant Sci. Lett.,* 22, 39, 1981.
31. **Potts, J. R. M., Weklych, R., and Conn, E. E.,** The 4-hydroxylation of cinnamic acid by *Sorghum* microsomes and the requirement for cytochrome P-450, *J. Biol. Chem.,* 249, 5019, 1974.
32. **Kjellbom, P. and Larsson, C.,** Preparation and polypeptide composition of chlorophyll-free plasma membranes from leaves of light grown spinach and barley, *Physiol. Plant.,* 62, 501, 1984.
33. **Lord, J., Kagawa, T., Moore, T. S., and Beevers, H.,** Endoplasmic reticulum as the site of lecithin formation in castor bean endosperm, *J. Cell. Biol.,* 57, 659, 1973.
34. **Ninnemann, H., Strasser, R. J., and Butler, W. L.,** The superoxide anion as electron donor to the mitochondrial electron transport chain, *Photochem. Photobiol.,* 26, 41, 1977.

35. **Schmidt, W., Hart, J., Filner, P., and Poff, K. L.,** Specific inhibition of phototropism in corn seedlings, *Plant Physiol.,* 60, 736, 1977.

36. **Leong, T-Y. and Briggs, W. R.,** Evidence from studies with acifluorfen for participation of a flavin-cytochrome complex in blue light photoreception for phototropism of oat coleoptiles, *Plant Physiol.,* 70, 875, 1982.

37. **Widell, S. and Sundqvist, C.,** Fluorescence properties of plasma membranes from oats and cauliflower, unpublished results, 1984.

38. **Klemm, E. and Ninnemann, H.,** Correlation between absorbance changes and a physiological response induced by blue light in *Neurospora, Photochem. Photobiol.,* 28, 227, 1978.

39. **Klemm, E. and Ninnemann, H.,** Nitrate reductase — a key enzyme in blue light promoted conidiation and absorbance change of *Neurospora, Photochem. Photobiol.,* 29, 629, 1979.

40. **Ninnemann, H. and Klemm-Wolframm, E.,** Blue light controlled conidiation and absorbance changes in *Neurospora* are mediated by nitrate reductase, in *The Blue Light Syndrome,* Senger, H., Ed., Springer-Verlag, Berlin, 1980, 238.

41. **Poff, K. L., Loomis, W. F., and Butler, W. L.,** Isolation and purification of the photoreceptor pigment associated with phototaxis in *Dictyostelium discoideum, J. Biol. Chem.,* 249, 2164, 1974.

42. **Häder, D.-P. and Poff, K.,** Light induced accumulations of *Dictyostelium discoideum* amoebae, *Photochem. Photobiol.,* 29, 1157, 1979.

43. **Mitzka-Snabel, U. and Rau, W.,** Subcellular site of carotenoid biosynthesis in *Neurospora crassa, Phytochemistry,* 20, 63, 1981.

44. **Mancinelli, A. L.,** Photoregulation of anthocyanin synthesis. VIII. Effect of light pretreatments, *Plant Physiol.,* 75, 447, 1984.

45. **Benveniste, I., Salaun, J.-P., and Durst, F.,** Phytochrome mediated regulation of a monooxygenase hydroxylating cinnamic acid in etiolated pea seedlings, *Phytochemistry,* 17, 359, 1978.

46. **Galland, P. and Lipson, E. D.,** Photophysiology of *Phycomyces blakesleeanus, Photochem. Photobiol.,* 40, 795, 1984.

47. **Ninnemann, H.,** The nitrate reductase system, in *Blue Light Effects in Biological Systems,* Senger, H., Ed., Springer-Verlag, Berlin, 1984, 95.

Chapter 7

PLASMA MEMBRANE PURIFICATION

S. Widell and C. Larsson

TABLE OF CONTENTS

I. Introduction 100

II. Experimental Procedure 100
    A. Chemicals 100
        1. Dextran T 500, 20% (w/w) 100
        2. Polyethylene Glycol (PEG) 3350, 40% (w/w) 100
        3. Sucrose, Salts, and Buffers 100
    B. Preparation Procedure 101
    C. Possible Problems 101
    D. Membrane Marker Assays 105
        1. Cytochrome c Oxidase 105
        2. NADPH-Dependent Cytochrome c Reductase 105
        3. Light-Induced Absorbance Change (LIAC) 105
        4. K^+-Mg^{2+}-ATPase 105
        5. Glucan Synthetase II (GS II) 105

III. Concluding Remarks 105

Acknowledgments 107

References 107

I. INTRODUCTION

The small amount of plasma membrane vesicles present in a crude microsomal fraction can readily be separated from the vast majority of intracellular membranes by partition in an aqueous polymer two-phase system.[1] This method separates membrane vesicles according to differences in their surface properties,[2] rather than by size and density as does centrifugation. By phase partition, plasma membrane preparations of very high purity (estimated 95 to 99% from specific staining with silicotungstic acid) have been obtained from several species and organs.[3-6] Apart from the high purity, phase partition offers several other advantages as compared with gradient centrifugation: the method is rapid — plasma membranes are obtained within 2 to 4 hr from homogenization. A constant osmotic potential may be used throughout the procedure. The plasma membrane vesicles are of uniform sidedness, right side out, and are closed.[7] The yield is high — 50 to 80%. The preparation is easily applied to larger samples, and no sophisticated equipment is needed.

II. EXPERIMENTAL PROCEDURE

A. Chemicals

Dextran T 500 (average molecular weight 500,000) is purchased from Pharmacia Fine Chemicals, Uppsala, Sweden; and polyethylene glycol 3350 (average mol wt 3350) from Union Carbide in New York. (Note that polyethylene glycol 3350 was earlier designated polyethylene glycol 4000.) The following stock solutions are prepared.

1. Dextran T 500, 20% (w/w)

Dextran powder normally contains 5 to 10% water, and therefore the exact concentration of the solution should be determined. Layer 220 g of dextran on 780 g of water in a large Erlenmeyer flask. Shake thoroughly until no big lumps are visible. Heat on a water bath with gentle stirring until all dextran is dissolved. Transfer approximately 5 g of dextran solution to a preweighed 25-mℓ measuring flask, record the exact weight of the solution, and dilute to 25 mℓ. Make three replicates. Measure the optical rotation (due to the glucose units of the dextran molecules) with a polarimeter at 589 nm. The specific rotation is 199 degree mℓ g^{-1} dm^{-1} and the concentration is obtained from the following relation:

$$\text{Concentration}(\%,\text{w/w}) = \frac{\text{optical rotation} \times 25(\text{m}\ell) \times 100}{199 \times (\text{g}) \times \text{path length (dm)}}$$

Adjust to 20.0% (w/w) with water. If a polarimeter is not available, the dextran should be stored dry and relatively large stock solutions prepared (assuming, e.g., 5% water) to ensure reproducible results.

2. Polyethylene Glycol (PEG) 3350, 40% (w/w)

400 g of PEG 3350 is dissolved in 600 g of water.

3. Sucrose, Salts, and Buffers

Prepare these 4 to 20 times stronger than the final concentrations used. The desired pH of buffers is obtained by mixing appropriate amounts of KH_2PO_4 and K_2HPO_4 (potassium phosphate buffer), or by titrating H_2SO_4 with Tris (SO_4-Tris buffer). Do not adjust the pH with HCl, since this adds an uncontrolled amount of Cl$^-$, an ion which strongly affects the partition of membrane vesicles (see Figure 2).

The solutions may be stored in the cold room for a short period or otherwise be stored frozen (in airtight containers).

B. Preparation Procedure

The preparation of plasma membranes from cauliflower inflorescences is chosen as an example.[7,8]

A bulk phase system (240 g), which will provide the fresh upper and lower phases used in the batch procedure (Figure 1), is prepared in a separating funnel (Table 1). After temperature equilibration in the cold room, the phase system is well mixed and allowed to settle overnight. The upper and lower phases are collected and stored separately in the cold or frozen. A phase mixture of 19.0 g is prepared in a centrifuge tube of suitable size (Table 1).

Cauliflower inflorescences (70 g) are homogenized with an Omnimixer for 3×20 sec in 180 mℓ of 25 mM MOPS-NaOH, 0.25 M sucrose, 3 mM Na-EDTA, 0.1 mM MgCl$_2$, 8 mM cysteine, pH 7.8. The homogenate is filtered through a 120-μm nylon cloth and centrifuged at 10,000 g for 15 min. A microsomal pellet is obtained from the supernatant by centrifugation at 50,000 g for 45 min. This pellet is suspended in 5.5 mℓ of 0.25 M sucrose, 4 mM KCl, 5 mM potassium phosphate, pH 7.8, and 5.0 g of the suspension is added to the 19.0-g phase mixture to yield a 24.0-g phase system (Table 1). The phase system is thoroughly mixed and centrifuged for 5 min at 1000 g in a swinging bucket rotor to facilitate phase settling. Most (90 to 95%) of the upper phase (containing the plasma membranes) is carefully removed with a Pasteur pipette without disturbing the interface. The upper phase is added to a tube containing fresh lower phase obtained from the bulk phase system (Table 2); and the upper phase is repartitioned to increase the purity of the plasma membranes. The exact volume of the lower phase is not critical since the two phases are in equilibrium. The original lower phase, containing the bulk of intracellular membranes is similarly repartitioned with fresh upper phase. These repartition steps may be repeated once more to further increase separation. This batch procedure is summarized in Figure 1. By this procedure a highly purified plasma membrane preparation (U$_3$) is obtained, as well as a fraction of intracellular membranes (L$_3$) depleted in plasma membranes.

The final upper phase (U$_3$) is diluted severalfold with a suitable medium and the plasma membranes are pelleted by centrifugation at 100,000 g for 1 hr. Similarly, the final lower phase (L$_3$) and what is left from the microsomal fraction (MF) are diluted at least 10-fold and the membranes pelleted. The pellets are suspended in a medium suitable for the assays planned.

C. Possible Problems

The composition of the homogenization medium is not very critical, and may be varied to meet any particular need. However, the composition of the phase system is critical. With both dextran and PEG, batch-to-batch variations occur, which affect the partition of the membranes. To get a good separation, 80 to 95% of the plasma membranes should partition in the upper phase, while only 5 to 20% of the intracellular membranes should do so. A good strategy is to start with the conditions given in Table 1. If the material partitions too much in the upper phase (e.g., a relatively high activity of cytochrome c oxidase is found in the upper phase), the polymer concentrations should be increased. If the material (including the plasma membranes!) partitions too much at the interface and in the lower phase, the polymer concentrations should be decreased. Optimal polymer concentrations are determined by partitioning the material in a series of phase systems with increasing polymer concentrations (Table 2, Figure 2) above or below the polymer concentrations used in the pilot experiment. For these experiments, phase systems with a total weight of 4.0 g are suitable. After mixing and phase settling, 90% of the upper phase is transferred to another test tube, and both phases are diluted to the same final volume. To determine the partition of different membranes, appropriate assays are performed on aliquots from the diluted phases.

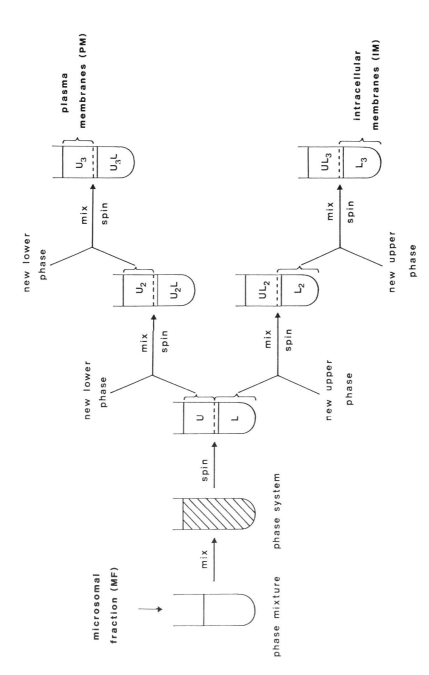

FIGURE 1.   Separation of plasma membranes from intracellular membranes by a batch procedure in three steps. Five g of microsomal fraction (10,000- to 50,000-g pellet) suspended in 0.25 M sucrose, 4 mM KCl, 5 mM potassium phosphate, pH 7.8, is added to a 19.0-g phase mixture to yield a 24.0-g phase system. The phase system is well mixed and allowed to settle. Phase settling is facilitated by centrifugation at 1000 g in a swinging bucket rotor for a few minutes. Then 90% of the upper phase (in Tube 1) is removed and repartitioned twice with fresh lower phase (Tubes 2 and 3). The lower phase + interface in Tube 1 is repartitioned twice with fresh upper phase. The fresh upper and lower phases are obtained from a bulk phase system (see Table 1).

Table 1
PHASE MIXTURE AND BULK PHASE SYSTEM
USED IN PLASMA MEMBRANE PURIFICATION

	Phase mixture (g)[a]	Phase system (g)[b]
Dextran, 20% (w/w)	7.80	78
PEG, 40% (w/w)	3.90	39
Buffer medium[c]	4.75	60
H₂O	+ 2.55	+ 63
Total weight	19.00	240
(sample)	+ 5.0	—
	24.0	

Note: The final concentrations in the phase system and the phase mixture after addition of sample will be 6.5% (w/w) polyethylene glycol 3350, 6.5% (w/w) Dextran T 500, 4 mM KCl, 5 mM potassium phosphate, pH 7.8.

[a] Weigh into a transparent centrifuge tube of 40 to 50 mℓ volume.
[b] Weigh into a separating funnel which after temperature equilibration is thoroughly shaken. The phases are left to settle overnight and are thereafter separated into upper and lower phases. The material around the interface is discarded.
[c] 1.0 M sucrose, 16 mM KCl, 20 mM potassium phosphate, pH 7.8.

Table 2
COMPOSITION OF 4.0-g PHASE SYSTEMS IN A
POLYMER SERIES

	Conc of dextran and PEG % (w/w)				
	5.7	5.9	6.1	6.3	6.5
Dextran, 20% (w/w) (g)	1.14	1.18	1.22	1.26	1.30
PEG 40% (w/w) (g)	0.57	0.59	0.61	0.63	0.65
Buffer medium[a] (g)	0.90	0.90	0.90	0.90	0.90
H₂O (g)	0.99	0.93	0.87	0.81	0.75
Sample[b] (mℓ)	0.40	0.40	0.40	0.40	0.40

Note: Final concentrations will be 5.7 to 6.5% of dextran and PEG, 4 mM KCl, 5 mM potassium phosphate, pH 7.8.

[a] 1.0 M sucrose, 16 mM KCl, 20 mM potassium phosphate, pH 7.8.
[b] In 0.25 M sucrose, 4 mM KCl, 5 mM potassium phosphate, pH 7.8.

Not only the polymer concentrations but also the ion composition of the phase system strongly affect the partition of membranes. This is because the ions give rise to an interfacial potential difference of a few millivolts in the phase system.[9] Divalent ions, such as HPO_4^{2-} and SO_4^{2-}, have a somewhat higher affinity for the lower phase and thus give a ''positively charged'' upper phase. Since all biological membranes are negatively charged at neutral pH, they will prefer the upper phase in a phase system with phosphate or sulfate buffer, provided the polymer concentrations are not too high (Figure 2). On the other hand, monovalent ions, such as Cl^-, have a somewhat higher affinity for the upper phase and therefore give phase systems with ''negatively charged'' upper phases. The inclusion of Cl^- in the phase systems thus strongly affects the partition of membranes (Figure 2), and may be used in addition to polymer concentrations to optimize separation (Table 3). Since several ions strongly affect

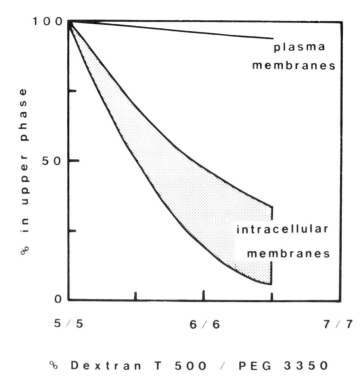

FIGURE 2.   Partition of plant membrane vesicles in a phase system buffered with 5 mM potassium phosphate, pH 7.8. When the polymer concentrations are increased, the material starts to collect at the interface. Different membrane vesicles, are, however, differently affected. Right-side-out plasma membrane vesicles and protoplasts, i.e., particles exposing the outer surface of the plasma membrane to the phase system, are scarcely affected, whereas intracellular membranes are excluded from the upper phase and partition at the interface + lower phase. Note that a similar effect is obtained when keeping the polymer concentrations constant (and relatively low) and adding chloride to a final concentration of 5 to 10 mM.

Table 3
COMPOSITION OF 4.0-g PHASE SYSTEMS IN A SALT SERIES

	KCl conc (mM)				
	0	2	4	6	8
Dextran 20% (w/w) (g)	1.14	1.14	1.14	1.14	1.14
PEG 40% (w/w) (g)	0.57	0.57	0.57	0.57	0.57
Buffer medium[a] (g)	0.90	0.90	0.90	0.90	0.90
KCl, 0.1 M (g)	0.00	0.08	0.16	0.24	0.32
H_2O (g)	0.99	0.91	0.83	0.75	0.67
Sample[b] (mℓ)	0.40	0.40	0.40	0.40	0.40

Note: Final concentrations of 5.7% (w/w) dextran, 5.7% (w/w) PEG, 5 mM potassium phosphate, pH 7.8, and 0 to 8 mM KCl.

[a]   1.0 M sucrose, 20 mM potassium phosphate, pH 7.8.
[b]   In 0.25 M sucrose, 5 mM potassium phosphate, pH 7.8.

the partition, it is important to consider the ions carried over to the phase system with the sample. The pellet should therefore be well drained before it is suspended in the medium used in the phase system. Note that the plasma membrane, in contrast to intracellular membranes, is relatively insensitive to both increased polymer and chloride concentrations (Figure 2). This makes it relatively easy to find a phase system which gives a good separation.

D. Membrane Marker Assays

1. Cytochrome c Oxidase

Cytochrome c oxidase[10] (marker for mitochondria) is assayed by measuring the oxidation of reduced cytochrome c at 550 nm. Cytochrome c is reduced with a few crystals of dithionite prior to assay. The final assay medium should contain 15 μM cytochrome c, about 50 mM phosphate buffer, pH 7.5, 0.01% Triton® X-100, and sample. The reaction is initiated by adding the appropriate amount of sample (10 to 40 $\mu \ell$) to 1 mℓ of assay medium. The difference in the rate of change of A_{550} with respect to time before and after sample addition is measured and the activity calculated using an extinction coefficient of $21 \cdot 10^3 \, M^{-1} \, cm^{-1}$. The final value is expressed as nmol min $^{-1}$.

2. NADPH-Dependent Cytochrome c Reductase

NADPH-dependent cytochrome c reductase[10] (marker for the endoplasmic reticulum, but activities are also associated with the plasma membrane[6,8]) is assayed by measuring the reduction of oxidized cytochrome c at 550 nm. The final assay medium should contain 30 μM cytochrome c, about 50 mM phosphate buffer, pH 7.5, 0.01% Triton® X-100, 2 mM KCN (in order to inhibit cytochrome c oxidase), 0.1 mM NADPH, and 10 to 40 $\mu \ell$ sample in a total volume of 1 mℓ. The reaction is initiated by the addition of NADPH. The baseline is calculated from the difference in the rate of change of A_{550} with respect to time after and before NADPH addition in an assay system without NADPH. The activity is expressed in nmol min^{-1} by use of the extinction coefficient of $21 \cdot 10^3 \, M^{-1} \, cm^{-1}$.

3. Light-Induced Absorbance Change (LIAC)

LIAC (light induced absorbance change, Δ (A_{428} to A_{410}), supposed to be a marker for the plasma membrane if used in combination with a high affinity of the material for the upper phase[5]) is induced by 30 sec to 1 min of red light (about 15 W · m^{-2}) in the presence of methylene blue (2 to 20 μM) as photoreceptor. Sample (about 0.1 mg protein in 1 mℓ of buffer with a neutral pH containing about 3 mM EDTA) is added to a cuvette, followed by methylene blue (in a negligible volume). The measurements are performed with a dual-wavelength spectrophotometer, operating in the dual-beam mode. Irradiation will lead to absorbance changes (at most 0.01) related to a light-induced methylene blue-mediated reduction of a b-type cytochrome located in the plasma membrane. The degree of light reduction is expressed as Δ (ΔA) (units of volume or protein)$^{-1}$. Make sure that the signal size is proportional to the amount of protein (something which should be checked in all marker assays).

4. K⁺-Mg²⁺ATPase

K$^+$-stimulated, Mg^{2+}-dependent ATPase (supposed to be a marker for the plasma membrane[10]) is assayed by measuring the liberation of P_i. The final assay medium should contain 1 mM ATP, 1 mM MgSO$_4$, 30 mM Tris-Mes, pH 6.0, and be with or without 25 mM KCl, in a final volume (sample included) of 0.5 mℓ. The reaction is initiated by addition of 50 $\mu \ell$ sample (5 to 25 μg protein) and the test tubes are incubated at 38°C for 15 to 30 min. The reaction is stopped with TCA (50 $\mu \ell$ 33%), and the test tubes are placed on ice. To each tube the following is added: 2 mℓ 0.2% SDS, 1.5 mℓ ammonium molybdate (3.57 mM in 0.8 M H$_2$SO$_4$), and 1 mℓ SnCl$_2$-solution (17.2 mM ascorbic acid, 2.22 mM SnCl$_2$

in 0.3 M H$_2$SO$_4$). The latter solution should be prepared immediately before use. A standard is obtained by mixing 2.5 mℓ 0.02 mM KH$_2$PO$_4$, 1.5 mℓ ammonium molybdate, and 1 mℓ SnCl$_2$ solution. A blank is obtained by mixing 2.5 mℓ H$_2$O, 1.5mℓ ammonium molybdate, and 1 mℓ SnCl$_2$-solution.

The absorbance at 680 nm is measured after 60 to 180 min. The K$^+$-stimulated part of the activity is obtained from the difference between activities in the presence of MgSO$_4$ + KCl and in the presence of only MgSO$_4$. The plasma-membrane-associated activity is latent[7] and a low concentration of detergent (e.g., 0.02% Triton X-100) in the assay medium may therefore stimulate the activity severalfold, and is recommended.

5. Glucan Synthetase II (GS II)

Glucan synthetase II (supposed to be a plasma membrane marker but also present in intracellular membranes) is assayed with a high UDP-glucose concentration and no Mg^{2+}. GS I, which is thought to be a Golgi marker, is in contrast assayed with a low UDP-glucose concentration and in the presence of Mg^{2+}. The following ingredients are added to poly-propylene tubes: 20 $\mu\ell$ H$_2$O and 40 $\mu\ell$ incubation buffer (see below), and the tubes are placed in a water bath at 25°C. After 5 min, 100 $\mu\ell$ of sample (20 to 40 μg protein) is added, and the contents mixed and incubated 10 to 25 min. The reaction is terminated by addition of 4 mℓ ice-cold ethanol (70%), and 200 $\mu\ell$ 0.1 M MgCl$_2$ and 160 $\mu\ell$ boiled microsomal fraction are added. After mixing, the tubes are placed in a boiling water bath for 2 min. They are then covered and left overnight at 0 to 4°C. The samples are washed on Whatman® GF/F filters (prewashed three times by swirling in a beaker with distilled water and/or ethanol) five times with 4 mℓ 70% ethanol. Then the filters are transferred to scintillation vials and left to dry at 60°C. Then 5 mℓ of scintillation fluid is added to each vial and the β-radiation is measured for 2 to 10 min, depending on the activity present. One control, ''nonspecific labeling'', is run exactly as above, with addition of buffer instead of sample. An internal standard of the counting efficiency is run by adding a known amount of ^{3}H-UDP-glucose (e.g., half of what is added for the samples) to filters which have been processed as the sample filters, and placed in scintillation vials prior to adding the ^{3}H-UDP-glucose.

For the incubation buffer, 1.19 g cellobiose is added to 9 mℓ of 50 mM Tris acetate, pH 8.0, and dissolved by heating. The solution is cooled to room temperature and 200 $\mu\ell$ of a solution containing 28.9 mg UDP-glucose/mℓ 50 mM Tris-acetate, pH 8.0 is added. To 1.1 mℓ of this solution, add 9 $\mu\ell$ ^{3}H-UDP-glucose (Amersham, 9 Ci/mmol, 1 μCi/ℓ) and 167 $\mu\ell$ distilled water. The concentrations in the samples during incubation will then be as follows: UDP-glucose: 0.218 mM, cellobiose: 0.15 M, and Tris-acetate: 11 mM.

Note that the pH optimum for the plasma-membrane-associated activity should be checked for each plant material,[7] to increase the specificity of the assay for the plasma-membrane-bound GS II. The activity is latent,[7] and the inclusion of a low concentration of detergent may therefore stimulate the activity severalfold. A problem is that most detergents also seem to inhibit the enzyme.[7]

III. CONCLUDING REMARKS

Phase partition produces a highly purified plasma membrane preparation (U$_3$), as well as a fraction of intracellular membranes depleted in plasma membranes (L$_3$). These two fractions are suitable for determining whether a reaction resides in the plasma membrane or in the intracellular membranes, or in both.

Phase partition procedures have been developed for the fractionation of intracellular mem-branes, including intact chloroplasts[11] and intact leaf mitochondria,[12] and for subfractionation of thylakoid membranes, including inside-out vesicles.[13] For the endoplasmic reticulum,

Golgi apparatus, and tonoplast, no method based on phase partition has been developed so far. However, these membranes may be obtained from some tissues with reasonable purity by sucrose gradient centrifugation.[14-16]

ACKNOWLEDGMENTS

We wish to thank Professor T. Murphy, Department of Botany, University of California, Davis, California, for important comments on the manuscript as well as for linguistic help. We also thank Dr. Per Kjellbom, Department of Biochemistry, University of Lund, for valuable suggestions on the assays.

REFERENCES

1. **Larsson, C.,** Plasma membranes, in *Modern Methods of Plant Analysis,* New Series, Vol. 1, *Cell Components,* Linskens, H. F. and Jackson, J. F., Eds., Springer-Verlag, Berlin, 1985, 85.
2. **Albertsson, P.-A.,** *Partition of Cell Particles and Macromolecules,* 3rd ed., John Wiley & Sons, New York, 1986.
3. **Widell, S. and Larsson, C.,** Separation of presumptive plasma membranes from mitochondria by partition in an aqueous polymer two phase system, *Physiol. Plant.,* 51, 368, 1981.
4. **Lundborg, T., Widell, S., and Larsson, C.,** Distribution of ATPases in wheat root membranes separated by phase partition, *Physiol. Plant.,* 52, 89, 1981.
5. **Widell, S., Lundborg, T., and Larsson, C.,** Plasma membrane from oats prepared by partition in an aqueous polymer two phase system. On the use of light induced cytochrome *b* reduction as a marker for the plasma membrane, *Plant Physiol.,* 70, 1429, 1982.
6. **Kjellbom, P. and Larsson, C.,** Preparation and polypeptide composition of chlorophyll-free plasma membranes from leaves of light-grown spinach and barley, *Physiol. Plant.,* 62, 501, 1984.
7. **Larsson, C., Kjellbom, P., Widell, S., and Lundborg, T.,** Sidedness of plant plasma membrane vesicles purified by partitioning in aqueous polymer two-phase systems, *FEBS Lett.,* 171, 271, 1984.
8. **Widell, S. and Larsson, C.,** Distribution of cytochrome *b* photoreductions mediated by endogenous photosensitizer or methylene blue in fractions from corn and cauliflower, *Physiol. Plant.,* 57, 196, 1983.
9. **Johansson, G.,** Partition of salts and their effects on partition of proteins in a dextran-poly(ethylene glycol)-water two-phase system, *Biochim. Biophys. Acta,* 221, 387, 1970.
10. **Hodges, T. K. and Leonard, R. T.,** Purification of plasma membrane bound adenosine-triphosphatase from plant roots, *Meth. Enzymol.,* 32, 392, 1974.
11. **Larsson, C., Collin, C., and Albertsson, P. A.,** Characterization of three classes of chloroplasts obtained by counter-current distribution, *Biochim. Biophys. Acta,* 245, 425, 1971.
12. **Gardeström, P., Eriksson, I., and Larsson, C.,** Preparation of mitochondria from green leaves of spinach by differential centrifugation and phase partition, *Plant Sci. Lett.,* 13, 231, 1978.
13. **Andersson, B. and Akerlund, H.-E.,** Inside-out membrane vesicles isolated from spinach thylakoids, *Biochim. Biophys. Acta,* 503, 462, 1978.
14. **Wagner, G.,** Higher plant vacuoles, in *Isolation of Membranes and Organelles from Plant Cells,* Hall, J. L. and Moore, A. L., Eds., Academic Press, London, 1983, 83.
15. **Lord, J. M.,** Endoplasmic reticulum and ribosomes, in *Isolation of Membranes and Organelles from Plant Cells,* Hall, J. L. and Moore, A. L., Eds., Academic Press, London, 1983, 119.
16. **Green, J. R.,** The Golgi apparatus, in *Isolation of Membranes and Organelles from Plant Cells,* Hall, J. L. and Moore, A. L., Eds., Academic Press, London, 1983, 135.

Ecological Relevance of Blue Light Effects

Chapter 8

CELLULAR AND FUNCTIONAL PROPERTIES OF THE STOMATAL RESPONSE TO BLUE LIGHT

Eduardo Zeiger

TABLE OF CONTENTS

I.   Introduction 112

II.   Kinetic Properties of the Stomatal Response to Blue Light 112

III.   A Kinetic Model of the Blue Light Response 114

IV.   Blue Light Induces Proton Extrusion from Guard Cell Protoplasts 117

V.   Mechanistic and Functional Implications of the Stomatal Response to Blue Light 117

Acknowledgments 119

References 119

I. INTRODUCTION

Stomata sense many endogenous and external stimuli, including light, intercellular CO_2 concentrations, temperature, relative humidity, phytohormones, and air pollutants.[1] Stomatal apertures, which modulate the gas exchange of leaves and other aereal organs of plants, are determined by the integrated response of guard cells to all these stimuli. An adequate understanding of stomatal movements therefore requires the characterization of metabolic events mediating signal transduction and turgor regulation in guard cells. Because of the dominant role of light in the modulation of stomatal responses, the precise elucidation of photoreception and its coupling with guard cell metabolism is of central interest for plant physiologists and cellular biologists.

The light responses of stomata are complex and depend on the operation of three different photoreceptors in the guard cells — chlorophyll,[2] a blue light photoreceptor,[3] and phytochrome —[4,5] and light-dependent interactions between the stomata and the photosynthesizing mesophyll.[1] The specific blue light response of stomata is the main subject of this paper; detailed analyses of the other light responses and their interactions have been published elsewhere.[1,6] The reader is also referred to a recent review on blue light and stomatal function for an overview of the subject and a phenomenological description of the interaction of the blue light photoreceptor and the other photosystems in the guard cells.[3] This chapter concentrates on recent progress which has enhanced our understanding of the properties of the blue light photosystem in stomatal guard cells.

II. KINETIC PROPERTIES OF THE STOMATAL RESPONSE TO BLUE LIGHT

Despite the unambiguous demonstration of a specific blue light response of stomata,[1] the detailed photobiological properties of that response have been elusive because of intrinsic difficulties in their characterization. One problem has been the requirement for a clear separation between the specific stomatal response to blue light and the stomatal responses mediated by the blue light responses of guard-cell chloroplasts and the underlying mesophyll tissue. An additional difficulty is that of quantification. A classical means of separating the specific light responses of stomata from that of the mesophyll are experiments with stomata in epidermal peels, which characterize changes in stomatal apertures as a function of light. These measurements, however, are usually obtained at discrete intervals, with substantial variability within and between samples. Gas-exchange experiments, in which the responses of stomata in an intact leaf are monitored as changes in the relative humidity in the leaf chamber, can provide continuous, real-time measurements with high resolution, but the results have been hard to interpret because they represent the combined response of stomata and the mesophyll.

Some of these problems were recently overcome in gas-exchange experiments with leaves of *Commelina communis,* the dayflower, using a double beam protocol combining continuous, high-intensity, red light with short (1 to 100 sec) pulses of blue light. High-intensity red light is used to saturate the photosynthetic responses of the guard-cell chloroplasts and of the mesophyll tissue.[7,9] Those conditions made it possible to study the specific, blue light-dependent, stomatal opening in the intact leaf, in response to low-intensity, continuous blue light.[3] That protocol was recently improved using pulsed blue light, which has been utilized with other experimental systems,[10] to obtain a quantitative analysis of the stomatal response to blue light.

A typical response of a *Commelina* leaf to a 30-sec pulse of blue light (250 μmol m^{-2} sec^{-1}) under a background of continuous red light (500 μmol m^{-2} sec^{-1}) is shown in Figure 1. Stomatal conductance increased in response to a blue light pulse, reaching a maximum

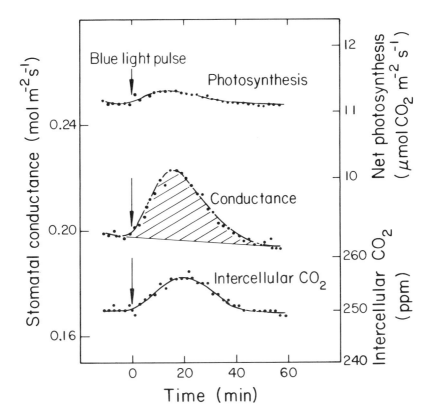

FIGURE 1. Profile of the increase in stomatal conductance following a 30-sec pulse of blue light (250 μmol m^{-2} sec^{-1}, arrow), given in background of approximately 500 μmol m^{-2} sec^{-1} red light. Two attached leaves of *Commelina communis* were enclosed in a gas-exchange cuvette (Armstrong Enterprises, Palo Alto, Calif.) Stomatal conductance and net photosynthesis were calculated from measurements of leaf transpiration and net CO_2 uptake.[10,11] Intercellular CO_2 concentrations were calculated from the conductance and photosynthetic rate measurements. Data were taken every 2 min with a computerized data acquisition system. Leaves were continuously irradiated with broad-band, red light supplied by a 300-W Sylvania (PAR56/2MFL) lamp, filtered through a wide-band hot mirror (OCLI, Santa Rosa, Calif.), a Kodak 1A filter, and a layer of No. 5A Cinemoid. The blue light pulses were given normal to the adaxial surface of the leaves, using a General Electric Gemini 300 projection lamp, filtered through a OCLI wide-band hot mirror, a 5-cm-thick water bath, and a Roehm and Haas 2424 Plexiglass filter. Light fluence rates were measured with a Li-Cor quantum meter. The conductance increase in response to a blue light pulse were quantified by integrating the area under the conductance curve, above the baseline levels measured under continuous red light (shaded area in the figure). (Modified from Ino, M., Ogawa, T., and Zeiger, E., *PNAS, 82,* 8019, 1985; *Photochem. Photobiol.,* 42, 759, 1985.

within 15 min and returning to baseline levels within 50 to 60 min after the pulse.[10,11] The small increase in photosynthetic rate was clearly attributable to the higher intercellular CO_2 concentrations ensuing from stomatal opening and was therefore independent of any direct light response of the photosynthetic apparatus. Additional evidence for a strict dependency of the conductance changes on the blue-light-dependent photosystem of the guard cells is the observation that red light pulses, of up to 500 μmol m^{-2} sec^{-1} had no effect.

Similar observations in several species, including soybean *(Glycine max),* sugarcane *(Saccharum* spp.), the orchid *Paphiopedilum harrisianum,* broad bean, *(Vicia faba* L.),[12] and bean *(Phaseolus vulgaris* L.),[13] showed that the stomatal capacity to respond to blue light pulses is not restricted to *Commelina* but appears to be present in both monocots and dicots.

Stomatal responses to blue light pulses were quantified in *Commelina* by integrating the area under the conductance curves above the baseline levels seen under continuous red light (shaded area of Figure 1). Figure 2A shows that increasing pulse duration (photon flux density: 250 μmol m^{-2} sec^{-1}) up to 30 sec caused higher responses, with no further increases observed with pulse durations longer than 30 sec. A plot of these responses relative to that observed after a 50-sec, saturating pulse showed approximately first-order kinetics, with a half time of about 9 sec (Figure 2B).

Commelina leaves responded to consecutive, saturating pulses 50 min apart, with seemingly identical kinetics, with no indication of an adaption to previous pulses.[14] On the other hand, the observation showing that the response to a single pulse saturated after a certain pulse duration, indicated that the restoration of the response capacity was time dependent. The kinetics of that recovery was studied with a protocol involving two 50-sec, saturating pulses separated by increasing time intervals (Figure 3A).[10,11] As with the response to single pulses, the responses to two pulses were quantified by integrating the area under the conductance curves above baseline levels, and plotted relative to the total area of a response to two consecutive pulses separated by a 50-min interval (Figure 3B). This response was also suggestive of first-order kinetics, with a half time of approximately 9 min.[10,11]

III. A KINETIC MODEL OF THE BLUE LIGHT RESPONSE

The observed properties of the responses to blue light pulses can be accounted for by a kinetic model postulating that one step of the phototransduction process includes a component which can exist in two interconvertible forms, *A*, and *B*.[10] *A* is an inactive form which is converted to *B* as a result of the photoexcitation of the blue light photoreceptor. *B* is an active form, with stomatal conductance assumed to be proportional to its concentration. *B* is converted back to *A* in a thermal reaction, with the complete process represented by the reaction

$$A \underset{k_d}{\overset{k_e}{\rightleftharpoons}} B$$

where k_e and k_d are rate constants for the light and dark (thermal) reactions, respectively.

On the basis of this model, we interpreted the single-pulse experiments in terms of the formation of some amount of *B*. With time, after the pulse *B* is converted back to *A*. Since k_e is significantly larger than k_d, the amount of *B* produced by a blue light pulse would be expected to show an apparent exponential saturation kinetics, as a function of pulse duration. It is also apparent from the model that the total amount of *B* formed by two saturating pulses increases with time between pulses, to a maximum of twice the amount of *B* formed by a single pulse. The amount of *B* formed by a second pulse would depend on the amount of *A* available at the time of the pulse, that is, on the amount of *B* converted back to *A*, from the time of the first pulse. It therefore follows that the total area under the conductance curve, which is assumed to be proportional to the integrated amount of *B*, would increase with time between pulses, as observed. (For a more detailed discussion of this model see Iino et al.[10])

A further, important prediction of the model is that, under continuous blue light, prevailing photostationary levels of *B* would depend on photon flux densities and the rate constants of the light and dark reactions. A comparison of the calculated, steady-state levels of *B* predicted by the model, with the steady-state levels of stomatal conductance which were measured at different photon flux densities of blue light, supported the hypothesis that blue light-induced increases in stomatal conductance are proportioned to the photostationary levels of *B*.[10]

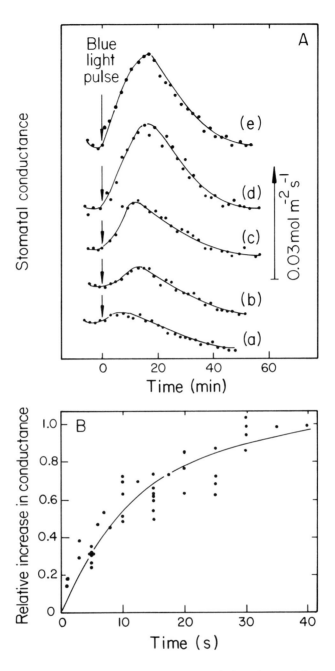

FIGURE 2.   Kinetics of the stomatal response in intact leaves of *Commelina communis* to a single pulse of blue light (arrows) in a background of continuous red irradiation. (A) Relationship between the increases in stomatal conductance and pulse duration. (a) 1, (b) 3, (c) 10, (d) 30, and (e) 100 sec. Photon fluences of background red light, blue light, and other experimental conditions as in Figure 1. (B) Plot of the conductance increases as a function of pulse duration, relative to the response to a saturating (50 sec) blue-light pulse. (Modified from Zeiger, E., Iino, M., and Ogawa, T., *Photochem. Photobiol.*, 42, 759, 1985.

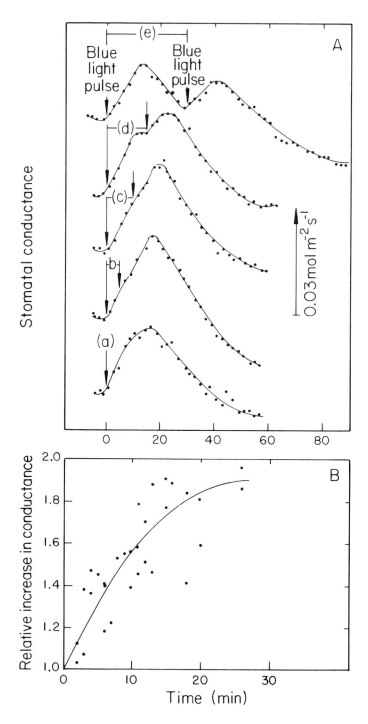

FIGURE 3.   Kinetics of the stomatal responses in intact leaves of *Commelina communis* to two 50-sec blue-light pulses. (A) Relationship between conductance increases and time between pulses. Pulses given as indicated by arrows. (a) single pulse, (b) 5-, (c) 10-, (d) 15-, and (e) 30-min interval. The slight decrease in baseline levels over time was also observed in the absence of blue light pulses and was most likely a result of slow decline in conductance ensuing from daily rhythms. Photon fluences and other experimental conditions as in Figure 1. (B) Plot of conductance increases as a function of time between pulses, relative to the response to two pulses 50 min apart. (Modified from Iino, M., Ogawa, T., and Zeiger, E., *PNAS,* 82, 8019, 1985; *Photochem. Photobiol.,* 42, 759, 1985.

The model does not specify the identify of the *A* and *B* forms. *A* and *B* could be inter-convertible forms of the photoreceptor itself, redox reactants coupled to the photoreceptor, or a "high-energy" state like a proton gradient. Some speculations on the possible identity of the *A* and *B* forms are presented below.

IV. BLUE LIGHT INDUCES PROTON EXTRUSION FROM GUARD-CELL PROTOPLASTS

It is well established that stomatal movements result from turgor changes in guard cells, which ensue from variations in osmotic potentials mediated by active ion transport at the guard-cell membranes. Ion uptake is most likely driven by proton extrusion, which generates an electrochemical gradient sustaining passive fluxes of K^+ and Cl^-.[1] It is therefore apparent that during light-dependent stomatal opening, photoreception needs to be transduced into ion fluxes and turgor buildup.

Guard-cell protoplasts, isolated by enzymatic digestion, respond to blue light by swelling, in a K^+-dependent response.[15] These observations were interpreted as a blue light-dependent stimulation of K^+ uptake leading to a higher osmotic potential of the protoplasts. In the absence of a cell wall, these higher osmotic potentials led to a significant volume increase.[15]

The relationship between photoreception and the cellular events underlying stomatal movement was further analyzed in experiments characterizing the changes in the pH of suspensions of guard cell protoplasts, following blue light pulses applied in a background of red light.[16] Blue light pulses consistently elicited a rapid acidification of the medium (Figure 4) which declined after 8 to 10 min. Like stomatal opening in the intact leaf, acidification could be elicited again after 30 min, with pulses given at shorter time intervals being progressively less effective. Acidification of the medium in response to a blue light pulse was completely inhibited by 50 μM DES or 10 μM CCCP, and was insensitive to DCMU.[16]

The observed acidification could be caused by CO_2 released into the medium because of a blue light-mediated enhancement of respiration.[17] Higher CO_2 release by respiring cells would partially convert to bicarbonate and H^+, causing acidification. Since the pH of the suspension medium was 6.2, about half of the CO_2 released would convert to bicarbonate and H^+. Alternatively, the observed acidification could be caused by blue light-induced proton extrusion.[3] These alternatives were tested by measuring rates of acidification at pH 5.3, 6.2, and 6.8. A CO_2-dependent acidification would be markedly pH dependent, with higher rates of acidification expected at the higher pH. In contrast, the observed acidification was essentially pH-independent within the measured range, pointing to proton extrusion as the underlying mechanism.[16] This conclusion is supported by the inhibition of acidification by CCCP. This proton translocator is an effective uncoupler of respiration and has been reported to double the rates of oxygen uptake in guard cell protoplasts.[18] In contrast, 10 μM CCCP completely inhibited the blue light-dependent acidification. We therefore conclude that blue light mediates proton extrusion in guard cells and that this response is the basis of the observed stomatal opening in the intact leaf.

V. MECHANISTIC AND FUNCTIONAL IMPLICATIONS OF THE STOMATAL RESPONSE TO BLUE LIGHT

The recent findings on the kinetics of the blue-light response and the established relationship between blue light photoreception in guard cell protoplasts and proton extrusion, have mechanistic and functional implications.

The correlation between the blue light-dependent increases in stomatal conductances and the predicted, relative concentrations of B[10] substantiates a kinetic model postulating a phototransduction process in which the photostationary levels of B determine the extent of

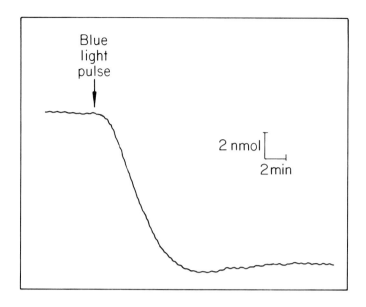

FIGURE 4.   Blue light-dependent acidification in a suspension of guard cell protoplasts from *Vicia faba* L. irradiated with background red light (800 μmol m^{-2} sec^{-1}); 30-sec blue light pulse (10 μmol m^{-2} sec^{-1}) given at the time indicated by arrow. Protoplasts were isolated as described in Reference 18. Suspension medium had 0.35 *M* mannitol, 0.5 m*M* MES-NaOH (pH 6.2), 10 m*M* KCl. and 1 m*M* CaCl$_2$. Red light supplied by a 500-W tungsten lamp (DAY DAK, Sylvania) filtered by a Corning 2-61; blue light was supplied by a 300 W-lamp (EXR, Sylvania), filtered by a Corning 5-60 filter and a 0.2% copper sulfate solution. About 10^6 protoplasts in a 1.2 mℓ suspension medium were continuously stirred in a plexiglass chamber, at 25°C. Measurements were obtained with a Radiometer pH meter and a S410A Beckman glass electrode and plotted in a E&K chart recorder. The figure was traced from the chart record.[16]

stomatal opening in response to blue light. The findings showing that blue light pulses cause proton extrusion by guard cell protoplasts suggest that the generation of a proton gradient driving ion uptake is a component of that blue light-dependent process.

The mechanism mediating the generation of the proton gradient remains to be elucidated. Proton extrusion could result from the activation of a plasmalemma ATPase or a membrane-bound, electron transport system (see Figure 3 in Reference 3 for a detailed description of these possible mechanisms). Further studies are also needed to determine if *B* is an interconvertible form of the photoreceptor itself, or a metabolic intermediate that is either a component of an electron transport system or a modulator of the activity of the ATPase, perhaps through a phosphorylation step.

The observation that a pulse of blue light generates a response which persists for minutes after the stimulus indicates that proton extrusion or stomatal opening do not depend on a continuous supply of photochemical energy from the blue quanta. It is possible that energy stored during the irradiation is subsequently utilized to drive the response, but that possibility implies an unusually high apparent quantum efficiency of the response, which, for proton extrusion, was estimated to be approximately tenfold higher than that for ATP synthesis by guard cell chloroplasts.[16] Alternatively, the phototransduction process could rely on an energy source independent of the blue light photosystem, like photophosphorylation or oxidative phosphorylation.[8]

The extent of the blue light-dependent enhancement of stomatal conductance above baseline levels observed under high-intensity red light provides additional information on the relative contributions of the two photoreceptor systems of guard cells to light-dependent stomatal opening.[7,8] In *Commelina* leaves irradiated with high-intensity red light and continuous blue light expected to yield maximal photostationary levels of *B*, the blue light-dependent conductance increases were consistently less than 20% of the values seen under red light alone,[10,11] indicating that, in the intact leaf, the blue light response is not the largest component driving light-dependent stomatal movements. At moderate to high photon fluences, the light response appears to depend primarily on photophosphorylation by guard cell chloroplasts.[7,19]

The blue light response shows, nevertheless, some specific functional correlates. Increases in the photostationary steady-state levels of *B* ensuing from rising photon flux densities appear to play a major role in the stomatal opening observed in dark to light transitions, like the blue light-dependent opening at dawn.[20] Of additional interest is the likelihood of a role of the blue light response in the reported kinetics of stomatal conductance changes in response to sun flecks in leaves growing in deep shade.[21] In these conditions, stomata open in response to sun flecks and remain in the open condition for several minutes after the end of the irradiation by the sun fleck. These kinetics are consistent with a role of the blue light response in the modulation of stomatal conductance in shaded leaves exposed to sun flecks, which deserves investigation. Finally, metabolic events like the modulation of malate synthesis[9] appear to be important in the regulation of guard cell metabolism by blue light.

The characterization of conductance increases in response to blue light pulses in several species indicates that the general features of the blue light photosystem observed with *Commelina* are likely to be widespread among higher plants. On the other hand, it will be of significant interest to establish whether the observed kinetic properties of *Commelina* are the same in different species, how it varies with developmental stages and habitats, and what, if any, are the contributions of the blue light-dependent photosystem to leaf adaptation and acclimation.

The emerging information on the properties of the blue light photosystem of stomata underscores its usefulness in the elucidation of basic properties of the phototransduction events involving blue light. The clear localization of the blue light response in a single cell type, its specific correlation with ion transport and proton extrusion, and its connection with relevant ecophysiological properties of the leaf make the stomatal response a valuable model system. The availability of specific kinetic parameters at the leaf and cellular levels is likely to accelarate the pace of future progress.

ACKNOWLEDGMENTS

The research reported in this paper has been supported by grants of the National Science Foundation, the Department of Energy, and the U.S. Department of Agriculture. I thank Dr. D. Grantz for valuable comments and criticisms.

REFERENCES

1. **Zeiger, E.,** The biology of stomatal guard cells, *Annu. Rev. Plant Physiol.,* 34, 441, 1983.
2. **Zeiger, E., Armond, P., and Melis, A.,** Fluorescence properties of guard cell chloroplasts, *Plant Physiol.,* 67, 17, 1981.
3. **Zeiger, E.,** Blue light and stomatal function, in *Blue Light Effects in Biological Systems,* Senger, H., Ed., Springer-Verlag, Berlin, 1984, 484.
4. **Roth-Bejerano, N. and Itai, Ch.,** Involvement of phytochrome in stomatal movement, *Physiol. Plant.,* 52, 201, 1981.

5. **Holmes, M. G. and Klein, W. H.,** Phytochrome regulates endogenous circadian rhythms in stomatal movements, *Plant Physiol.,* 75S, 130, 1984.

6. **Sharkey, T. D. and Ogawa, T.,** Stomatal responses to light, in *Stomatal Function,* Zeiger, E., Farquhar, G., and Cowan, I., Eds., Stanford University Press, Stanford, 1985, in press.

7. **Zeiger, E. and Field, C.,** Photocontrol of the functional coupling between photosynthesis and stomatal conductance in the intact leaf, *Plant Physiol.,* 70, 370, 1982.

8. **Schwartz, A. and Zeiger, E.,** Metabolic energy for stomatal opening. Roles of photophosphorylation and oxidative phosphorylation, *Planta,* 161, 129, 1984.

9. **Ogawa, T., Ishikawa, H., Shimada, K., and Shibata, K.,** Synergistic action of red and blue light and action spectra for malate formation in guard cells of *Vicia faba* L., *Planta,* 142, 61, 1978.

10. **Iino, M., Ogawa, T., and Zeiger, E.,** Kinetics of the stomatal response to blue light: a minimum kinetic model, *PNAS,* 82, 8019, 1985.

11. **Zeiger, E., Iino, M., and Ogawa, T.,** The blue light response and some mechanistic implications, *Photochem. Photobiol.,* 42, 759, 1985.

12. **Grantz, D. and Zeiger, E.,** in preparation.

13. **Larque-Saavedra, A. and Zeiger, E.,** in preparation.

14. **Lipson, E. D., Galland, P., and Pollockm, J. A.,** Blue light photoreceptors in *Phycomices* investigated by action spectroscopy, fluorescence lifetime spectroscopy and two-dimentional gel electrophoresis, in *Blue Light Effects in Biological Systems,* Senger, H., Ed., Springer-Verlag, Berlin, 1984, 228.

15. **Zeiger, E. and Hepler, P. K.,** Light and stomatal function: blue light stimulates swelling of guard cell protoplasts, *Science,* 196, 887, 1977.

16. **Shimazaki, K., Iino, M., and Zeiger, E.,** Blue light-dependent proton extrusions by guard-cell protoplasts of *Vicia faba, Nature (London),* 319, 324, 1986.

17. **Kowallik, W.,** Blue light effects on respiration, *Annu. Rev. Plant Physiol.,* 33, 51, 1982.

18. **Shimazaki, K., Gotow, K., Sakaki, T., and Kondo, N.,** High respiratory activity of guard cell protoplasts from *Vicia faba* L., *Plant Cell Physiol.,* 24, 1049, 1983.

19. **Shimazaki, K. and Zeiger, E.,** Cyclic and non-cyclic photophosphorylation in isolated guard cell chloroplasts from *Vicia faba* L., *Plant Physiol.,* 1985, in press.

20. **Zeiger, E., Field, C., and Mooney, H. A.,** Stomatal opening at dawn: possible roles of the blue light response in nature, in *Plants and the Daylight Spectrum,* Smith, H., Ed., Academic Press, New York, 1981, 391.

21. **Pearcy, R. W. and Calkin, H.,** Carbon dioxide exchange of C_3 and C_4 tree species in the understory of a Hawaiian forest, *Oecologia,* 58, 26, 1983.

Chapter 9

MARINE PLANTS AND BLUE LIGHT

Matthew J. Dring

TABLE OF CONTENTS

I.   Introduction 122

II.   Blue Light Environment of Marine Plants 122

III. Types of Response to Blue Light 124
      A.   Photoorientation Responses 125
            1.   Phototropism 125
            2.   Phototaxis 125
            3.   Induction of Polarity 126
            4.   Chromatophore Displacement 126
      B.   Metabolic Responses 127
            1.   Protein and Carbohydrate Metabolism 127
            2.   Enzyme Activity 127
            3.   Pigment Content 127
      C.   Developmental Responses 128
            1.   Vegetative Development 129
            2.   Reproductive Development 130
      D.   Photoperiodic Responses 132

IV.   Physiological Aspects of the Blue Light Responses in Marine
      Algae 133
      A.   Action Spectra 133
      B.   Sensitivity of Responses to Blue Light 134

V.   Regulation of Plant Growth in the Sea by Blue Light 135

References 138

I. INTRODUCTION

The "deep blue sea" is a rather hackneyed phrase, but one which encapsulates the popular idea that the marine environment is a blue light environment. In such an environment, it could be argued, the blue light responses discussed in this book must take on a special significance. The habitats occupied by marine plants are indeed often (but not always) rich in blue wavelengths, and marine plants as a group are certainly rich in blue light responses, but we should not jump immediately to the attractive conclusion that marine plants are therefore "adapted" to their light environment. Many marine habitats are not blue, but white or green or even orange in color, while the widespread occurrence of blue light responses among marine plants may be a reflection of their phylogenetic status, rather than an adaptation to their environmental situation. Over 90% of the plant species found in the sea are algae. This is an extremely diverse and rather artificial group of plants, but its members are generally regarded as more "primitive" than the vascular plants which dominate the terrestrial flora. If blue light responses are typical of lower plants, and are superceded by phytochrome responses only higher up the evolutionary ladder, marine plants might be expected to display blue light responses even in a red sea! In addition, because marine plants are more diverse — both morphologically and physiologically — than plants from other habitats, we might expect to find types of blue light response among marine plants which are unknown outside this group. This chapter examines the blue light responses of marine plants in the light of these ideas.

II. BLUE LIGHT ENVIRONMENT OF MARINE PLANTS

The absorption and scattering (collectively known as the attenuance) of light by pure water are minimal at wavelengths between 430 and 470 nm,[1] and light at these wavelengths can penetrate 250 m of pure water before its irradiance is reduced to 1% of the original value. The attenuance rises sharply at shorter and longer wavelengths so that an equivalent reduction in irradiance would occur within 30 m for ultraviolet (UV) radiation at 350 nm, and within 15 m for red radiation (>625 nm). Since dissolved sea salts have little influence on the optical properties of water, these predictions also apply to filtered sea water from deep unproductive oceans (e.g., Sargasso Sea, central North Pacific). However, as primary productivity increases in oceans with improved nutrient supplies (e.g., upwelling areas) and in shallower waters over the continental shelves or in coastal regions, the concentration of both plankton and inorganic detritus increases and reduces light penetration throughout the spectrum. The increased productivity and the larger amounts of runoff from the land also lead to an increasing concentration of dissolved organic compounds, which are characterized by their yellow color ("yellow substances" or "gilvin"[1]) and their high absorption of blue light. Consequently, the decrease in overall light penetration from open ocean to inshore waters is accompanied by a shift in the wavelength of maximal transmittance, from 465 nm in the clearest oceans to 575 nm in turbid coastal waters. This progression has been described quantitatively in an optical classification of sea water,[2] which defines three types of oceanic water (Types I to III) and nine types of coastal water (Types 1 to 9). Among both groups of water types, low numbers represent clear waters and high numbers the more turbid waters.

The actual spectrum experienced by marine plants in their natural habitat depends on the water type and the depth at which they are growing. Near the surface, the spectrum will not differ significantly from that of natural daylight (i.e., sun + skylight[3]), in which all wavebands are fairly evenly represented. With increasing depth in all water types, however, the spectrum will become increasingly narrow, and complete wavebands at either end of the visible spectrum (i.e., blue at <500 nm, or red at >600 nm) may be effectively eliminated. The depth at which this occurs, and which end of the spectrum is removed,

Table 1
DEPTH OF PHOTIC ZONE FOR
PHYTOPLANKTON AND
LAMINARIALES (1% DEPTH) AND
FOR MULTICELLULAR ALGAE
(0.05% DEPTH) IN DIFFERENT
WATER TYPES

	Depth (m)	
Water type	**1%**	**0.05%**
Oceanic		
I	105	175
II	55	95
III	32	55
Coastal		
1	27	48
3	17.5	31.5
5	11.5	20
7	8.0	14
9	6.0	10.5

Based on data from References 3 and 5.

depends upon the water type. At 10 m below the surface in most coastal waters, at least 70% of the total quanta are concentrated within a waveband of only 100 nm bandwidth, whereas in oceanic waters at least 30 m are needed to achieve the same degree of concentration.[4] The wavebands removed will be blue in coastal types 7 and 9, both blue and red in types 1 to 5, but red (or red and green) in the oceanic water types. The most extreme spectra to which marine plants are exposed will be those at the greatest depth at which plants grow in each water type. These depths have been found to correspond to the depths at which the total visible irradiance (350 to 700 nm) is reduced to 1% of its surface value (the ''1% depth'') for the deepest growing large brown algae (usually kelps) and phytoplankton, and to 0.05% of the surface value (the ''0.05% depth'') for the deepest macroscopic seaweeds (usually crustose coralline algae).[5] The actual depths at which these limits occur in each water type are listed in Table 1, and the spectral distributions of the light fields at these depths are shown in Figure 1.

Comparing light fields at the same ''optical depth'', there is clearly a similar degree of narrowing of the spectrum in all water types. At the 1% depth, about two thirds of the total quanta are restricted to a waveband of 75 nm bandwidth, whereas at the 0.05% depth, over 75% of the quanta are found within 75 nm (Figure 1). At the lower limit of the photic zone in all water types, therefore, plants are effectively growing in only one region of the visible spectrum. This is pure blue light for oceanic types I and II at depths of over 100 m, and it is here — but only here — that the ''deep blue sea'' really exists! More productive oceanic waters (type III) are blue-green in color, while the waters off most open coasts (types 1 to 5) are almost pure green. More turbid coastal waters, such as those found in enclosed seas (e.g., the North Sea) and in shallow bays and estuaries (types 7 and 9) tend to be yellow or orange. Blue wavelengths are absorbed rather rapidly in all coastal waters, and make up less than 10% of the total quanta at the 0.05% depth in Types 1 to 5, and less than 1% in types 7 and 9 (Figure 1). Even at the relatively shallow depth of 6 m in type 7 water, the mean daytime irradiance in summer in the 400 to 500 nm waveband was only 0.8 μmol m^{-2} sec^{-1} (or 0.05 mol m^{-2} d^{-1}), compared with a total visible irradiance (400 to 700 nm) of 15.0 μmol m^{-2} sec^{-1} (or 0.82 mol m^{-2} d^{-1}).[7] Far from being characteristically blue,

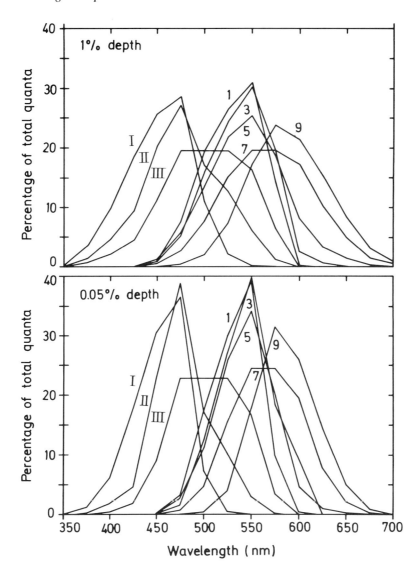

FIGURE 1. Spectral distribution of underwater light at the 1% and the 0.05% depth (see Table 1) in all water types, expressed as a percentage of the total visible quanta (350 to 700 nm) in each 25-nm waveband. (Calculated from data in Reference 2. From Dring, M. J. and Lüning, K., in *Encyclopedia of Plant Physiology*, New Series, Vol. 16B, Shropshire, W. and Mohr, H., Eds., Springer-Verlag, Berlin, 1983, 545. With permission.)

therefore, many habitats occupied by marine plants are markedly deficient in blue light. It is the absence of blue wavelengths, rather than their presence, which may create an ecological dimension for some blue light responses of marine plants.

III. TYPES OF RESPONSE TO BLUE LIGHT

This section reviews all of the blue light responses reported for marine plants, but some of the descriptions are very brief because many responses are similar in type to the blue light responses of other plants described elsewhere in this book. A few of the responses described here are, indeed, the principal examples of certain types of response, and are featured as such in other parts of the book, but out of their specifically marine context.

More detailed descriptions are reserved for responses which are unknown in other plant groups, or ones which are referred to in the subsequent discussion of the physiology or ecology of blue light responses among marine plants.

A. Photoorientation Responses

This group of responses includes all those in which a plant reacts to a directional or localized light stimulus by growth, movement, or rearrangement of cellular organelles.

1. Phototropism

There were many early observations of phototropic responses in seaweeds (see review by Banbury[8]), but few of these responses have been studied in detail. It was not until the mid 1970s that any evidence was obtained to confirm that phototropism in red and brown algae was indeed a blue light response — as in all higher plants and fungi. Although there is still no detailed action spectrum for phototropism in any alga, it has now been shown that blue is the active waveband for the positive phototropic response of the filaments of the red alga *Polysiphonia*,[9] and for the negative phototropic responses of the rhizoids of the red algae *Griffithsia*[10] and *Polysiphonia*,[9] the conchocelis filaments of *Porphyra*,[9] the primary filaments of the sporelings of the brown alga *Scytosiphon*,[11] and the haptera of the kelp *Alaria*.[12] Phototropism is clearly a common symptom of the blue light syndrome in marine algae, and it would be interesting to know just how closely the physiology of these responses in red and brown algae resembles that of the better known phototropic responses of higher plants and fungi.

2. Phototaxis

Phototaxis is just one of the several types of photomovement found in motile microorganisms, which are discussed by Häder in Volume 1, Chapter 9. Most microscopic plants are aquatic organisms, and a large proportion of these are marine. Consequently, several of the stars of phototactic research are marine species. There are four main groups of photosynthetic flagellates in the sea: the green flagellates (including members of the Volvocales and the Prasinophyta); the dinoflagellates; the Chrysophyta (including silicoflagellates and coccolithophorids); and the Cryptophyta. Phototaxis has been studied in several green flagellates from the sea (*Dunaliella, Platymonas, Stephanoptera*[13,14]) and in several dinoflagellates (*Peridinium, Gonyaulax, Prorocentrum;*[13] *Gymnodinium*[15]), but no marine representatives of the other two groups have been shown to exhibit phototactic behavior. In addition to these microscopic species, both brown and green seaweeds may have flagellate stages in their life history, and these may also be either positively or negatively phototactic.[16] *Ulva* gametes are positively phototactic before fusion, but the quadriflagellate zygotes which are formed by the fusion of two gametes are negatively phototactic.[17] The action spectrum for the positive response of the gametes peaks in the blue,[13] and resembles the action spectra for other green flagellates and most dinoflagellates,[13,15] but no action spectrum is available for the negative response of the zygotes. The phototactic behavior of the flagellate gametes or zoospores of brown algae has not been systematically studied.

Plants of the Rhodophyta (red algae) and Cyanophyta (blue-green algae or cyanobacteria) have no flagellated stages, but many unicellular or filamentous forms or spores of macroscopic species exhibit gliding motion and may be phototactic. The only phototactic marine species to have been studied in detail is the unicellular red alga *Porphyridium cruentum*.[18] The action spectrum for phototaxis in this organism shows a sharp peak at 443 nm, with shoulders at 416 and 470 nm, but no response in the near-UV region of the spectrum — or in the green region, which is absorbed strongly by its main photosynthetic pigment, phycoerythrin. The lack of response to green light is surprising only because other phycoerythrin-containing species (the blue-green *Phormidium* spp. and the cryptophytan flagellate *Cryp-*

tomonas) exhibit action spectra for phototaxis which implicate phycoerythrin in photoreception.[18] Even more surprising is another marine flagellate, the dinoflagallate *Prorocentrum,* which responds phototactically only to green light (480 to 640 nm, with a peak at 580 nm[13]), even though it does not possess phycoerythrin, or any other known pigment with a comparable absorption spectrum.

There appears to be nothing to distinguish phototaxis in marine plants from that in other organisms, but this is not unexpected since related species of unicellular plants are frequently found in marine, freshwater, and subaerial habitats. However, it is clearly something of an oversimplification to regard phototaxis purely as a blue-light response, and detailed studies should be concentrated on the smaller and phylogenetically more isolated groups to establish a broad base for generalizations.

3. Induction of Polarity

In many lower plants, newly formed spores or newly fertilized eggs are completely apolar, but the position of the first outgrowth or the plane of the first cell division determines where the apex and where the base of the plant will be formed. Several environmental factors may influence the direction of development, but a light gradient across the cell is particularly effective. The usual pattern in photosynthetic plants is for a rhizoid to develop on the shaded side of a zygote or spore, whereas in fungi the germ tube develops on the illuminated side.[19] The large zygotes (80 to 100 μm in diameter) of the brown alga *Fucus* and related genera (e.g., *Pelvetia, Ascophyllum*) provide a particularly convenient system for studies of cell polarity, and much of our knowledge of how light affects this process is based on work with *Fucus.* The action spectrum for the response peaks in the visible at 455 nm, although 254 nm radiation is even more effective than blue light.[20] Blue light has also been shown to be the most effective waveband for inducing polarity in another fucoid alga, *Cystoseira barbata,*[21] and in the siphonaceous green alga *Codium fragile.*[22] The very much smaller zoospores of the green alga *Ulva mutabilis* also put out a rhizoid towards the dark pole of the cell when they germinate in a horizontally directed white light field,[23] but the wavelength dependence of this response has not been studied. It seems probable that the spores of red algae would show similar responses, and some investigation of this aspect of germination in the Rhodophyta would help us to build up a complete picture of blue light responses in this group.

4. Chromatophore Displacement

The chromatophores of many brown seaweeds found on rocky shores respond to strong illumination by migrating to the side walls of the cell and thus increasing the overall transmittance of the thallus.[24] This response resembles that observed in many green plants from freshwater habitats (including green algae, mosses, pteridophytes, and angiosperms),[25] and appears similarly to be controlled by blue wavelengths.[26] It is curious, therefore, that green and red seaweeds from the intertidal zone do not react to strong light in the same way.[24] In green algae, at least, this is not because the chromatophores are immobile. Those of *Ulva lactuca* display a marked circadian rhythm, migrating to the side walls during the day and to the top and bottom walls at night,[27] but this pattern is not influenced by irradiance. Many other green algae in the sea are siphonaceous species, which possess neither cell walls nor a flat thallus, so that chromatophore orientation with respect to light direction is more difficult to assess. However, the local irradiation of parts of the thallus of *Bryopsis* and *Acetabularia* with spots of light results in the accumulation of chromatophores in the illuminated area, and blue light is again the most effective waveband.[28-30] Since the chromatophores are being continually moved about in the streaming cytoplasm of these species, the accumulation in a light spot could be caused by photoinhibition of the streaming, which is observed in *Bryopsis,* but active phototactic movement of the chromatophores towards the light appears to occur in *Acetabularia.*

B. Metabolic Responses
1. Protein and Carbohydrate Metabolism

A detailed account of the effects of blue light on carbohydrate metabolism is given by Kowallik in Chapter 2 of Volume 1, and it is only necessary here to point out that several marine species from widely different taxonomic groups exhibit similar responses to flowering plants and microscopic algae. The giant unicells of *Acetabularia* have been most intensively studied, and the metabolic contrasts between blue- and red-grown cells include higher photosynthetic and dark respiration rates in blue light, and a massive accumulation of starch in red light.[31] The photosynthetic rate of the brown alga *Dictyota* has also been found to decrease in the absence of any blue wavelengths, and there is a similar accumulation of polysaccharides in red light.[32] Red algae tend to be rather neglected by biochemists, but the branching filaments of *Acrochaetium daviesii* responded to red and blue light in much the same way as the green and brown seaweeds. The carbohydrate to protein ratio was 1.7 in red light and only 0.75 in blue.[33] Finally, the planktonic members of the marine flora are represented by the green flagellate *Dunaliella,* in which the rates of dark respiration and of starch breakdown were stimulated by low irradiances of blue light, as compared with darkness, but equal irradiances of red light had a similar effect.[34] Other effects which appear to be induced by both blue and red light will be discussed later (see Section D, ''Photoperiodic Responses'').

2. Enzyme Activity

In two of the species discussed in the previous paragraph, the accumulation of polysaccharide in cells grown in red light can be attributed to a decline in the activity of enzymes involved in carbohydrate degradation. Pyruvate kinase and UDPG pyrophosphorylase from red-light grown *Acetabularia* had a lower specific activity than the same enzymes from blue-grown cells, but their activity increased rapidly when cells were transferred to blue light.[35] In *Acrochaetium,* it was glucose-6-phosphate dehydrogenase and 6-phosphogluconate dehydrogenase which were less active in red- than in blue-grown plants.[33] In neither of these species did there seem to be a direct regulatory effect of blue light on the enzymes involved. Blue light appeared to stimulate protein synthesis in general, and the increased specific activity of certain enzymes may simply be due to their rapid turnover, which causes them to disappear more quickly than other enzymes in red light.[35]

3. Pigment Content

When the freshwater green algae *Chlorella* and *Scenedesmus* were grown in blue light or in low irradiances of white light, the chlorophyll concentration of the cells increased and so did the ratio of chlorophyll b to chlorophyll a, whereas red light and high irradiances of white light resulted in the opposite response.[36,37] This mimicking of the effects of low irradiance by blue light, and of high irradiances by red light, is the reverse of the situation often observed in higher plants,[38] and has been interpreted as an adaptation to the aquatic environment, in which poorly lit habitats tend to be rich in blue wavelengths.[37] This generalization about light quality in deep water is more justified for oceanic waters than for coastal waters (see Section II) or most bodies of fresh water,[1] and there have been analogous suggestions that the low-intensity adaptations of the photosynthetic apparatus of deep oceanic phytoplankton are induced by blue light.[39,40] Unfortunately, few systematic studies of the influence of light quality on the pigment content of marine plants have been attempted, and the experimental evidence for a specific blue light response equivalent to low-intensity adaptation is much weaker for marine plants than for *Chlorella* and *Scenedesmus*.

A chrysophytan flagellate (*Ochromonas* sp.) collected from a deep oceanic habitat formed more chlorophyll in blue-green light than in red light,[40] but there was no evidence of an increase in light-harvesting chlorophylls relative to reaction center chlorophylls, as would be expected in low-light adapted cells. A cryptomonad from the same habitat (*Cryptomonas* sp.) showed no difference in chlorophyll content between blue and red light, although there was a slight increase in green light,[40] whereas chlorophylls a and c in the dinoflagellate *Prorocentrum* were 50% lower in blue light than in red, green, or white light.[41] The macroscopic red alga *Gracilaria* also had reduced chlorophyll-a concentrations after growth in blue light, compared with green or white light.[42] These conflicting and confusing results contrast with the evidence from a broad survey of marine phytoplankton,[43] which indicated that many species (8 out of the 18 examined) formed more chlorophyll in weak blue-green light than in the same irradiance of white light. Since the white light source contained some blue wavelengths, however, it is still not possible to conclude that blue light stimulates chlorophyll formation in marine plants.

This discussion of the effects of light quality on the pigment content of marine plants cannot be allowed to pass without some mention of the old controversy about "chromatic adaptation". Deep-sea plants have long been thought to be adapted to the blue or green light of their natural habitats by the possession of accessory photosynthetic pigments (e.g., fucoxanthin, peridinin, phycoerythrin) which absorb blue-green and green wavelengths. If this were true, we might expect the concentration of such pigments to be increased in blue and green light, and it is well known that in some blue-green algae (all species investigated have been from freshwater habitats), the balance between phycocyanin and phycoerythrin changes with light quality — phycoerythrin increasing in green light, and decreasing in red light.[44,45] This observation is often assumed to apply also to marine algae of all classes, but several recent investigations have shown that the pigment composition of most marine algae does not respond to changes in light quality. Seven species of seaweeds (including green, brown, and red algae) showed no significant differences in pigment ratios after growth in low irradiances of either natural daylight or green light,[46] and blue-green light had little influence on the relative amounts of different pigments in the wide range of marine phytoplankton studied by Vesk and Jeffrey.[43] Since it is the phycobilin pigments which respond to light quality in blue-green algae, other phycobilin-containing algae (i.e., red algae, cryptomonads) might be expected to be the best candidates for similar responses. This expectation has been disappointed by the red algae examined so far,[42,46] but the phycoerythrin content of *Cryptomonas* has been shown to increase rapidly in green light.[40] The characteristic accessory pigment of dinoflagellates, peridinin, was also found to increase markedly in green light in *Prorocentrum*.[41] The experimental evidence for chromatic adaptation in the sea seems to rest, therefore, on these two species of phytoplankton, and it appears to be green light, rather than blue light, which is effective in altering pigment composition on the rare occasions when light quality has any effect at all.

C. Developmental Responses

In order to detect specific effects of blue light on the development of plants, it is necessary to cultivate them for long periods in red light — or, at least in some light field which lacks blue wavelengths. For terrestrial plants, such treatments can be rightly criticized as extremely artificial, because there are no natural habitats which receive red light alone (or, indeed, blue light alone), and this experimental approach has rarely been used. In the sea, however, plants growing towards the lower limit of the photic zone will normally be exposed to light fields which lack either blue or red wavelengths, or both (see Figure 1), so that comparing the growth and development of marine plants in red and blue light fields is likely to provide useful information about their ecology as well as their physiology. Some of the species which have been examined in this way are reluctant to grow in red light. On transfer to red light, the growth of both *Acetabularia* and *Dictyota* declines to zero within 2 to 3 weeks,[32,35] presumably as a result of the metabolic and enzymatic changes which also occur in red light (see above). In such species, the inhibition of development in red light is probably an indirect

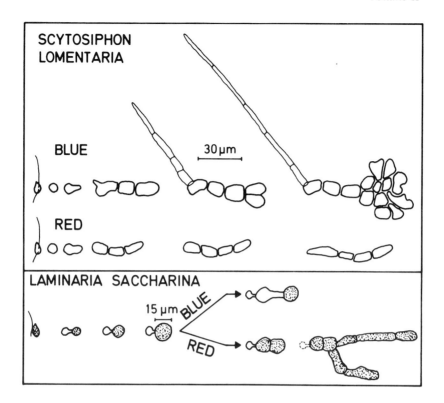

FIGURE 2. Morphological development of *Scytosiphon lomentaria* and female gametophytes of *Laminaria* spp. in blue (or white) and red light. (From Dring, M. J. and Lüning, K., in *Encyclopedia of Plant Physiology,* New Series, Vol. 16B, Shropshire, W. and Mohr, H., Eds., Springer-Verlag, Berlin, 1983, 545. With permission.)

consequence of the restricted growth. There are other species, however, which are capable of long-term growth in red light. The best studied examples are the microscopic gametophytes of *Laminaria* (which must be cultured in red light to prevent them from reproducing and dying; see below) and young plants of *Scytosiphon* but, unfortunately, we have no information about how red light affects the metabolism of these plants. At present, it is not even possible to say whether growth of most marine plants grinds to a halt in red light, like *Acetabularia* and *Dictyota,* or grow on indefinitely, like *Laminaria* and *Scytosiphon.* An investigation of a wide range of species would be extremely valuable to answer this question, and to establish whether the ability to grow in red light is related to the natural habitat of the species.

1. Vegetative Development

The most fundamental effect of blue light on the vegetative development of marine plants is seen in the brown algae *Scytosiphon* and *Petalonia.* Zoospores released by the erect phases of these species grow at similar rates in red and blue light, but develop quite differently. Plants in blue light grow into neat circular, two-dimensional crusts, which spread over and attach to the surface of the substrate, and develop long colorless hairs projecting into the water above. In red light, the plants remain entirely filamentous, branch irregularly in three dimensions, and develop neither hairs nor any attachment to the substrate.[47] The contrast between the developmental patterns in the two wavebands can be seen 4 to 5 days after germination, when the three- or four-celled filaments of blue-grown plants are already shorter and broader than red-grown plants, and the apical cell becomes heart shaped, prior to the first longitudinal division which signals the beginning of two-dimensional growth (Figure 2). The percentage of plants forming these so-called "heart cells" has been used as a measure of the effects of different light treatments in studies of the photobiology of this blue-light response.

The formation of hairs in *Scytosiphon* is very closely associated with the establishment of two-dimensional growth, and it has not yet proved possible to separate the two responses. This could mean that hair formation is simply a part of the developmental pattern leading to the formation of crusts, and should not be regarded as a distinct response to blue light. However, hairs are also formed following exposure to blue light in another brown alga, *Dictyota*,[32] and in the green alga *Acetabularia*,[35] so that this seems to be a typical response to blue light in marine plants. The arrangement of the hairs is quite different in each species. They occur singly in *Scytosiphon*, but in tufts in *Dictyota* (at densities of up to 150/cm² of thallus), and in apical whorls in *Acetabularia*. The physiological function of such hairs in marine plants has not been established, but the most attractive hypothesis (first proposed as long ago as 1892[48]) is that they act as nutrient antennae and assist in the absorption of inorganic ions from water outside the boundary layer surrounding the algal thallus. A rather fascinating speculative parallel can then be drawn between the stimulatory effects of blue light on hair formation in marine algae, and those on nitrate reductase activity in microscopic algae, fungi, and higher plants.[49]

2. Reproductive Development

The formation of caps, which signals the onset of reproduction in *Acetabularia*, occurs only in blue light,[50,51] but this may be a response to the better growth and protein synthesis of the plants in this light field, rather than a specific response to blue light.[35] In *Dictyota*, on the other hand, reproduction was observed only at wavelengths greater than 600 nm,[32] even though growth was reduced and polysaccharides accumulated under these conditions. It is possible that this reproductive response is a reaction to the poor growth and the increased carbon-to-nitrogen ratio, but more detailed investigation is clearly required, especially since egg release in the same species is stimulated by blue light (see below). The effects of blue light on reproduction in another group of brown algae, the kelps or Laminariales, are more clear-cut, and do not seem to be confounded with effects on growth. The primary cells of female gametophytes of *Laminaria saccharina* grew at the same rate in red and blue light for the first 8 days after settlement of the zoospores, but their subsequent development was quite different.[52] In blue light, the primary cell stopped growing and formed a single egg, which was extruded from the cell, leaving an empty oogonium (Figure 2). The life span of a female gametophyte in blue light at 15 C is thus completed in about 10 days. In red light, growth continues with the division of the primary cell and the development of a branched filamentous plant. Reproduction does not occur under these conditions, and vegetative growth may continue for years. Transferring a red-grown, filamentous gametophyte to blue (or white) light induces immediate reproduction, and many eggs will then be produced from a single gametophyte. Antheridium formation in the smaller male gametophytes is also induced by blue light.[53] Most of the detailed work on this response has been conducted with *Laminaria saccharina*, but the other European species of *Laminaria* and three species of kelps from California, including the giant kelp *Macrocystis pyrifera*, have been shown to exhibit a similar response to blue light.[53,54]

Although the formation of eggs in *Laminaria* gametophytes is stimulated by blue light, the subsequent release of these eggs from their oogonia may be inhibited (at least, for a time) by further blue light exposure. This response can be observed only in plants which have been induced to form eggs by a light/dark cycle of white or blue light during the first 8 days after settlement of the zoospores.[55] If this light/dark cycle is continued for the next 3 days, egg release will occur during the dark periods but not in the light. A similar time course of egg release is seen if the light/dark cycle is replaced by continuous darkness (Figure 3a), or continuous red or green light (Figure 3d and 3b), but continuous blue or white light

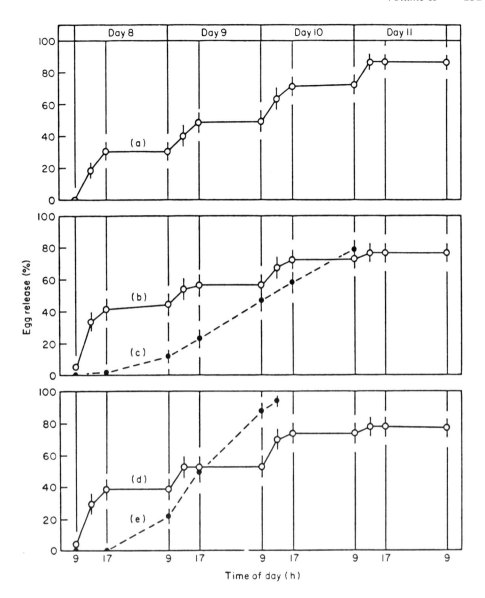

FIGURE 3. *Laminaria saccharina*. Time course of egg release in gametophytes transferred at beginning of Day 8 from a light/dark regime (16:8̄ hr into continuous conditions: (a) darkness; (b) green light — 541 nm; (c) blue light — 448 nm; (d) red light — 662 nm; (3) white light. (From Lüning, K., *Br. Phycol. J.*, 16, 379, 1981. With permission of the British Phycological Society.)

results in a very different pattern (Figure 3c and 3e). Release is completely inhibited at first, and then occurs gradually and continuously over the next 24 to 48 hr. These results suggest that the light/dark cycle which induced egg formation also activated a circadian rhythm controlling egg release. Since blue light clearly interferes with the expression of this rhythm, it seems likely that blue light is responsible for establishing it. Therefore, we may be dealing here with an effect of blue light on an endogenous rhythm (see Chapter 2), rather than a direct response of the egg-release mechanism to blue light.

The opposite conclusion seems to apply to the opposite effect of blue light on egg release in *Dictyota*. In this plant, the release of eggs is normally restricted to a 15-min period each day, which starts about 20 min after the beginning of the light period, even if this "light period" consists of only 1 min of very weak blue light.[56] Release will also occur every 23

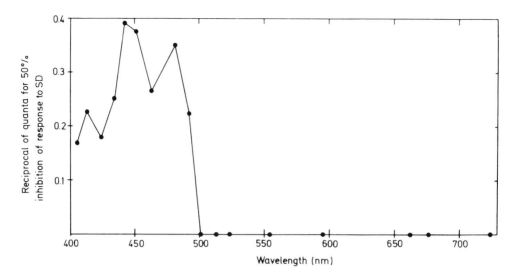

FIGURE 4.   *Scytosiphon lomentaria*. Action spectrum for the inhibitory effect of night breaks on the induction of erect thallus formation by SD at 12°C. (From Dring, M. J. and Lüning, K., *Planta*, 125, 25, 1975. With permission.)

to 24 hr, however, if the plants are grown in continuous light or continuous darkness.[57] Thus, the exact timing of egg release is determined by a pulse of blue light, provided that this is given during the correct phase of a circadian rhythm, but the rhythm persists in the absence of light pulses. The action spectrum for this response has peaks in the blue (at 464 nm) and near-UV (366 nm) regions, and thus is typical of blue light responses in many marine algae and other plants. However, the triggering of egg release in *Dictyota* is remarkable because it is both the most rapid response to blue light among marine plants, and the most sensitive (see below).

D. Photoperiodic Responses

All photoperiodic responses in flowering plants are thought to be mediated by phytochrome, because red wavelengths are most effective as night breaks or day extensions for both long-day (LD) and short-day (SD) plants, and the effects of red light can be reversed by far-red light.[58] A similar result was obtained with the first photoperiodic response to be investigated in an alga (the stimulation of conchospore formation in the red alga *Porphyra tenera*[59,60]), and this suggested that phytochrome was also associated with photoperiodism in lower plants. However, in the absence of a detailed action spectrum for the response in *Porphyra*, this conclusion could not be confirmed, and the subsequent investigations of photoperiodic responses in other marine algae have undermined it by showing that phytochrome is not involved in these species, and that photoperiodism in plants can be a blue-light response.

The most detailed action spectrum available refers to the SD response of a brown alga which has already figured prominently in this review, *Scytosiphon lomentaria*. The crusts, which develop in continuous white or blue light (see Section C.1), start to form erect tubular thalli on transfer to SD conditions.[61] As in most photoperiodic flowering plants, the response is inhibited by a short "night break" in the middle of a long dark period but, unlike all flowering plants, it is blue and not red light which is effective (Figure 4). It proved impossible to reverse the inhibitory effects of blue light by subsequent exposure to other wavelengths, so that there was no similarity at all to phytochrome action. Similar results were obtained when the photoperiodic response of the red alga *Acrosymphyton* was investigated.[62,63] In this species, the formation of tetraspores which occurred in SD was not inhibited by short night breaks, but only by 8-hr extensions of the main photoperiod with low-intensity blue (393 to 473 nm) light.

In two other photoperiodic species, the red alga *Rhodochorton* and the brown alga *Ascophyllum,* there is evidence that blue and red light are equally effective as night breaks,[64,65] although detailed action spectra are not available for either species. The inhibitory effects of red light on the SD response of *Rhodochorton* could not be reversed by far-red light,[64] so that phytochrome does not seem to be responsible for the red light effect. The detection of another "blue-and-red-light" response in the green flagellate *Dunaliella*[34] (see Section B.1) means that such responses have been reported in three algal groups with quite different photosynthetic pigments and phylogenetic origins. This wide distribution should provide a strong incentive to investigate the pigment or pigments responsible. Are these really blue light responses which are reinforced by the action of a second, independent, red-absorbing pigment, or are they mediated by a single pigment with an absorption spectrum similar to chlorophyll? Whatever the answer, it will be a novel one for students of photomorphogenesis in plants.

IV. PHYSIOLOGICAL ASPECTS OF THE BLUE LIGHT RESPONSES IN MARINE ALGAE

All that is needed to describe an effect of light on a plant as a "blue light response" is a demonstration that some aspect of the performance of the plant in broad-band blue light is markedly different from that in broad-band red light. This criterion is the only one to have been used in compiling the list of blue light responses for the review in the previous section, and little more than this is known about some of the responses. Most of the detailed information available about the other responses relates either to the action spectrum of the response, or to its sensitivity to blue light.

A. Action Spectra

True action spectra have been determined for only six of the responses described in Section III,[14,20,28,35,56,61] but effectiveness spectra are available for several other responses. Most of these spectra show a main peak at 440 to 450 nm, with subsidiary peaks (or shoulders) at 405 to 415 nm and at 480 to 490 nm, although some spectra (see, for example, References 13, 20, and 56) show a simpler structure in the 400- to 500-nm region, possibly because fewer wavelengths were investigated. Not all spectra cover the near-UV region, but many of those which do have a peak at 370 to 380 nm, often with low values at both 400 and 350 nm.[28,34,52,55,56] Many spectra relating to phototactic responses, however, lack a peak in the near-UV (see discussion by Nultsch[18]). Considering the wide range of response types and the taxonomic diversity of the plants in which they have been observed, there is a remarkable degree of agreement between the spectra, and most would be covered by the operational definition of "cryptochrome" as a blue-UV light photoreceptor "with peaks or shoulders at or around (370), 420, 450, and 480 nm."[66] It is not surprising that the green algae (i.e., *Acetabularia,*[35] *Bryopsis,*[28] *Dunaliella,*[13,34] *Platymonas,*[14] *Ulva*[13]) should fit so closely with this definition, since so much of our knowledge about blue light responses is based on work with other green algae (freshwater species originally, but often totally laboratory-bound now!) and higher plants. The similarity of the spectra obtained with four unrelated genera of brown algae (i.e., *Dictyota,*[56] *Fucus,*[20] *Laminaria,*[52,55] *Scytosiphon*[47,61]) suggests, however, that cryptochrome may also be of general occurrence in the Phaeophyta, and perhaps in other algal groups in the same evolutionary line (i.e., the "chromophytes" or brown-colored algae, including diatoms, chrysophytes, and dinoflagellates), although only the spectrum for *Gymnodinium*[15] has sufficient detail to support this for dinoflagellates. The red algae remain rather enigmatic. Neither of the two spectra available — for phototaxis

Table 2
PHOTON EXPOSURES REQUIRED FOR A 50% RESPONSE TO BLUE LIGHT IN SELECTED MARINE ALGAE

Genus	Response	Photon exposure	Ref.
a. High photon requirements			
Laminaria	Induction of egg formation	1.95 mol m^{-2}	52
Macrocystis	Induction of egg formation	2.60 mol m^{-2}	54
Scytosiphon	Induction of hair formation	1.97 mol m^{-2}	47
Scytosiphon	Induction of 2-D growth	2.25 mol m^{-2}	47
b. Medium photon requirements			
Laminaria	Inhibition of egg release	3.8 mmol m^{-2}	55
Rhodochorton	Inhibition of SD response by night break	0.6 mmol m^{-2}	64
Fucus	Photoinduction of polarity (primary response)	0.17 mmol m^{-2}	20
Scytosiphon	Inhibition of SD response by night break	0.02 mmol m^{-2}	61
Acetabularia	Induction of hair whorls	0.02 mmol m^{-2}	35
c. Low photon requirements			
Dictyota	Stimulation of egg release	0.08 μmol m^{-2}	56

in *Porphyridium*[18] and for photoperiodism in *Acrosymphyton*[62,63] — look like typical cryptochrome spectra, but neither are they sufficiently different to discount the presence of cryptochrome. Blue light has been shown to influence phototropism, metabolic responses, and photoperiodic control in red algae, and attention needs to be concentrated on these responses to establish whether the blue photoreceptor of red algae is similar to the cryptochromes of other plants.

B. Sensitivity of Responses to Blue Light

Different blue light responses in marine algae require quite different exposures to blue light in order to be activated. The total range of photon exposures which elicit a 50% response exceeds 7 orders of magnitude (Table 2), but much of this is accounted for by the contrast between a group of responses with particularly high photon requirements, and the single extremely sensitive response of *Dictyota*. The remaining responses cover only two orders of magnitude and are fairly typical of blue light responses in other plants in their sensitivity.

The threshold for the egg-release response in *Dictyota* at 464 nm occurs at about 6 nmol m^{-2},[56] which is not much higher than that for the most sensitive blue light response reported in higher plants, the first positive curvature of phototropism in oat coleoptiles (threshold at 458 nm, 0.7 nmol m^{-2}).[67] The slightly lower sensitivity of *Dictyota* could possibly be explained by the shading of the cryptochrome by photosynthetic pigments within the oogonium, and there may be a related explanation for the substantially reduced sensitivity of the algal responses with "medium" photon requirements (Table 2). However, the presence of different blue-light effects with medium and high photon requirements within the same species (e.g., *Scytosiphon, Laminaria;* Table 2) suggests that shading of the cryptochrome within the cell or thallus cannot explain why a group of vegetative and reproductive responses in brown algae should have such high photon requirements. In *Scytosiphon*, for example, different responses operate over a range of 10^5 in photon requirements, and this may indicate that different photoreceptors are involved in different responses.

The high light-requiring responses of these brown algae are also of particular interest because of the greatly extended period over which reciprocity has been shown to operate.

When *Scytosiphon* plants were exposed to a range of irradiances of blue light for various periods up to 96 hr, the percentage of plants initiating two-dimensional growth was proportional to the total photon exposure, irrespective of irradiation time.[7] Similarly, the fertility of female gametophytes of *Laminaria* after 3 or 4 weeks in various "white" light fields, some of which simulated spectral distributions underwater and all of which had a different proportion of blue light, was correlated with the total exposure to photons in the active waveband (i.e., 400 to 512 nm) and not with any other aspect of the irradiation treatments.[53] Experiments in which the morphogenetical development of *Laminaria* and *Scytosiphon* were followed at different depths in the sea[7] (see below) also indicated that these plants are able to count and respond to the total number of photons received over a period of 2 to 3 weeks. Such long-term reciprocity has not been reported in any other photobiological system, and it is very difficult to picture how the plants are able to count photons over a time period during which they are growing and altering their pigment content. The biochemical analysis of these blue light responses should make fascinating reading some day!

V. REGULATION OF PLANT GROWTH IN THE SEA BY BLUE LIGHT

Since marine plants are often found in blue light environments, it is tempting to conclude that the blue light responses of marine plants will be advantageous in their natural habitats. As so often with temptations, however, the attractions of this one are entirely superficial and disappear on closer examination. The incident light in deep oceanic waters contains a higher *proportion* of blue wavelengths than the light in terrestrial or shallow water habitats, but it does not contain more blue light in absolute terms — the irradiance at depth is always lower than nearer the surface. In most plants, blue light effects respond to the absolute amount of blue light received (up to a saturating level) and, except where there is an interaction with another pigment system (e.g., the phytochrome system; see Chapter 10 in Volume 1), the extent of the response to blue light is independent of other wavelengths present in the light field. Thus, the response to a mixture of wavelengths (including white light) is governed entirely by its blue light content. This means that deep oceanic sites will be *less* favorable (or, at best, no more favorable) for a blue light response than sites higher in the water column, and the existence of blue light responses in deep-sea plants cannot confer a specific advantage for life in the deep sea. On the contrary, blue light responses may be positively disadvantageous in deep coastal waters, because blue light decreases more rapidly with depth than other wavebands and may become limiting for a blue light response before the total visible irradiance becomes limiting for photosynthesis and growth. Some of the responses with high photon requirements (Table 2) have been shown to be inhibited by this mechanism in the field.

The morphological development of *Scytosiphon* plants, which were suspended in chambers at fixed depths in the sea off Helgoland, was followed for up to 4 weeks, and the time taken for 50% of the plants to develop a particular morphological feature was estimated.[7] The total exposure of the plants to both visible photons (400 to 700 nm) and blue photons during this period was then determined from the continuous records for irradiance meters suspended at the same depths. In April, it took nearly 18 days for *Scytosiphon* plants at 4 m to form hairs, whereas in May and June, 50% of the plants at the same depth formed hairs in about 6 days (Table 3). The total exposure of the plants to blue photons during these different lengths of time in the three experiments was 2 to 3 mol m^{-2} (similar to the photon requirement determined in the laboratory for this response; Table 2), but the total visible photon exposure in June was 2.5 times that in April and May, and the water temperature was different for all three experiments. Thus, hair production was more closely correlated with the incident number of blue photons than with any other factor, and this conclusion is supported by the results for plants at 6 m in the June experiment (Experiment III, Table 3). After 17.3 days,

Table 3
SCYTOSIPHON LOMENTARIA:
**HAIR FORMATION BY SPORELINGS IN THE SEA NEAR HELGOLAND
(NORTH SEA) IN RELATION TO LIGHT CLIMATE**

Experiment Date started	I April 25, 1973	II May 22, 1973	III June 28, 1974	III June 28, 1974
Temperature during experiment	6—8°C	10—11°C	14—15°C	14—15°C
Depth of sporelings (below MLWS)	4 m	4 m	3 m	6 m
Days to 50% response	17.8	5.9	6.0	17.3
				(5% response)
Incident light recorded during this period				
Photon exposure (mol m^{-2})				
Blue (400—500 nm)	2.12	1.74	2.78	0.92
White (400—700 nm)	11.04	10.06	25.60	14.13
Mean photon irradiance (μmol m^{-2} sec^{-1})				
Blue (400—500 nm)	2.3	6.5	11.0	0.8
White (400—700 nm)	12.2	29.5	81.0	15.0

Data from Reference 7.

these plants had been exposed to more visible photons than those at 4 m in April and May, and yet only 5% of the plants had formed hairs. The light at 6 m was predominantly green (water Type 7,with 75% of visible photons at 500 to 600 nm) and the total exposure to blue light was only 0.92 mol m^{-2} — about half of that required for a 50% response. The average visible irradiance (15 μmol m^{-2} sec^{-1}) was similar to that used to obtain optimal growth in the laboratory, and the plants at 6 m were larger than those which had formed hairs in the sea in April. The failure to form hairs cannot, therefore, be attributed to slow growth, and the low proportion of blue wavelengths in the light at this depth would seem to be the principal factor which has retarded their morphological development.

Similar experiments with *Laminaria* at Helgoland have indicated that during much of the winter (the season when zoospores are released by the sporophytes), there was insufficient light, even at only 2 m below low-tide mark, to support vegetative growth of the gameto-phytes.[53] The plants persisted as single cells, however, and reproduction of both male and female gametophytes occurred at a depth of 2 m as soon as the irradiance increased in February. At 5 m, the total exposure to blue quanta was well below that required for a 50% response (Table 2), and no reproduction was observed, even though the gametophytes reached a larger size than those at 2 m. Thus, the reproduction of *Laminaria* gametophytes at 5 m appeared to be delayed by the poor penetration of blue light through these waters, and this factor may, therefore, entirely prevent reproduction at greater depths. This could have an important influence on the distribution of the sporophyte generation, since fertilization and germination usually occur at the mouth of the oogonium. Similarly, the initiation of two-dimensional growth in *Scytosiphon* could be critical for the distribution of this species, because normal development results in a flat, circular crust, which attaches the plant to the surface of pebbles and rocks (see Section III.C.1).

There is also some circumstantial evidence for the limitation of the depth range of particular species of marine plants through blue light responses which control growth and phototaxis. As we have seen, both *Acetabularia* and *Dictyota* fail to grow in the absence of blue light (see Section III.C). *Acetabularia* is usually found in shallow tropical or subtropical waters in which there is unlikely to be a shortage of blue light, but *Dictyota* occurs in greener coastal waters on temperate shores. Since the species was only recorded down to a depth of 0.5 m in the type 7 waters around Helgoland,[68] whereas it is common and abundant at 20 m (i.e., below the 1% depth; see Table 1) in the bluer, type 3 waters off some Atlantic coasts of the British Isles,[69] it is possible that its blue light requirement for growth is responsible for the poor penetration of the species into Helgoland waters. More detailed studies of the depth range of *Dictyota* in different water types would be useful to test this idea. It would also be interesting to know whether seaweeds which are found at depth in turbid waters resemble *Laminaria* and *Scytosiphon* in their ability to grow in pure red light fields (see Section III.C.). It has also been suggested that the maximum depth at which certain dinoflagellates are found in coastal waters may be determined by the sensitivity of the phototactic mechanism of the cells to blue light.[15] The threshold irradiance for phototaxis in *Gymnodinium* was 0.1 W $\cdot$ m^{-2} at 453 nm, and this value is similar to that calculated for a depth of 15 m in type 5 coastal water. Since 15 m is close to the lowest depth from which nocturnal dinoflagellate populations have been reported, it seems possible that cells which migrate any deeper at night are unable to respond to light from the surface in the morning, and so never return to the photic zone.

All of these examples suggest that blue light responses of various types may prevent the growth of marine plants in the green or orange light fields which are typical of deep coastal waters. Of course, there will also be depths in clearer waters at which there would be too little blue light to activate such responses but, because there is little light at other wavelengths in these light fields, plant growth would probably be restricted by the shortage of light for photosynthesis at shallower depths than those at which the activity of blue light responses would be limited. It is for this reason that blue light responses are thought to be most relevant to the ecology of marine plants in coastal waters. Nevertheless, many species of marine plants are found only in the very blue light of deep oceanic waters, and phytoplankton populations are frequently most dense at depths of 90 to 130 m.[39] This suggests that the high proportion of blue wavelengths in the light fields at these depths favors the growth of these species, and we should ask what role blue light could have in this growth promotion.

Plants from deep oceanic habitats appear to be adapted to growth in low irradiances by the possession of high concentrations of photosynthetic pigments, and by the ability to increase this concentration as irradiance decreases. If marine plants were to use a blue-absorbing photoreceptor to measure the irradiance, we might expect chlorophyll synthesis to be inhibited by blue light, and this is observed in some species (see Section B.3). On the other hand, several phytoplankton species form more chlorophyll in blue-green than in white light,[43] which suggests that chlorophyll synthesis is inhibited by red light. It may, therefore, be a red-light response, rather than a blue-light response, which equips these plants to live in their blue light environment. An alternative strategy for deep-sea plants is to detect low irradiances by measuring the ratio of blue to red light. This would be analogous to the use by higher plants of the red to far-red ratio (measured by phytochrome) to detect shade,[70] and could be achieved either by a single pigment with two photoreversible forms (like phytochrome) or by two separate pigments working antagonistically. Although such a mechanism would detect low-light environments in oceanic and clear coastal waters, it would give the "wrong" answer in more turbid coastal waters, in which blue light is absorbed more rapibly than red, and the blue to red ratio decreases with depth.[71]

This discussion has necessarily been rather speculative, but it should clarify the reasons for regarding typical blue light responses as poor physiological mechanisms for adapting plants to the "deep blue sea". These speculations about how such adaptations could be achieved may, however, be useful "dummies" to be shot at by future investigators of the ecological relevance of blue light regulation in marine plants.

REFERENCES

1. **Kirk, J. T. O.,** *Light and Photosynthesis in Aquatic Ecosystems,* Cambridge University Press, Cambridge, 1983.
2. **Jerlov, N. G.,** *Marine Optics,* Elsevier, Amsterdam, 1976.
3. **Smith, H. and Morgan, D. C.,** The spectral characteristics of the visible radiation incident upon the surface of the earth, in *Plants and the Daylight Spectrum,* Smith, H., Ed., Academic Press, London, 1981, 3.
4. **Dring, M. J.,** Chromatic adaptation of photosynthesis in benthic marine algae: an examination of its ecological significance using a theoretical model, *Limnol. Oceanogr.,* 26, 271, 1981.
5. **Lüning, K. and Dring, M. J.,** Continuous underwater light measurements near Helgoland (North Sea) and its significance for characteristic light limits in the sublittoral region, *Helgol. Wiss. Meeresunters.,* 32, 403, 1979.
6. **Dring, M. J. and Lüning, K.,** Photomorphogenesis of marine macroalgae, in *Encyclopedia of Plant Physiology,* New Series, Vol. 16B, *Photomorphogenesis,* Shropshire, W. and Mohr, H., Eds., Springer-Verlag, Berlin, 1983, 545.
7. **Dring, M. J. and Lüning, K.,** Photomorphogenesis of brown algae in the laboratory and in the sea, *Int. Seaweed Symp.,* 8, 159, 1981.
8. **Banbury, G. H.,** Phototropism of lower plants, in *Encyclopedia of Plant Physiology,* Vol. 17, Part 1, *Physiology of Movements,* Ruhland, W., Ed., Springer-Verlag, Berlin, 1959, 530.
9. **Tussenbroek, B. I. van,** Effect of continuous unilateral irradiation on the conchocelis of *Porphyra umbilicalis* (L.) J. Ag. and some other red algae, *J. Exp. Mar. Biol. Ecol.,* 83, 263, 1984.
10. **Waaland, S. D., Nehlsen, W., and Waaland, J. R.,** Phototropism in a red alga, *Griffithsia pacifica, Plant Cell Physiol.,* 18, 603, 1977.
11. **Dring, M. J.,** unpublished data, 1974.
12. **Buggeln, R. G.,** Negative phototropism of the haptera of *Alaria esculenta* (Laminariales), *J. Phycol.,* 10, 80, 1974.
13. **Halldal, P.,** Action spectra of phototaxis and related problems in Volvocales, *Ulva*-gametes and Dinophyceae, *Physiol. Plant.,* 11, 118, 1958.
14. **Halldal, P.,** Ultraviolet action spectra of positive and negative phototaxis in *Platymonas subcordiformis, Physiol. Plant.,* 14, 133, 1961.
15. **Forward, R. B.,** Phototaxis by the dinoflagellate *Gymnodinium splendens* Lebour, *J. Protozool.,* 21, 312, 1974.
16. **Fritsch, F. E.,** *The Structure and Reproduction of the Algae,* Vol. 2, Cambridge University Press, Cambridge, 1945.
17. **Carter, N.,** An investigation into the cytology and biology of the Ulvaceae, *Ann. Bot.,* 40, 665, 1926.
18. **Nultsch, W.,** Effects of blue light on movement of microorganisms, in *The Blue Light Syndrome,* Senger, H., Ed., Springer-Verlag, Berlin, 1980, 38.
19. **Weisenseel, M. H.,** Induction of polarity, in *Encyclopedia of Plant Physiology,* New Series, Vol. 7, *Physiology of Movements,* Haupt, W. and Feinleib, M. E., Eds., Springer-Verlag, Berlin, 1979, 485.
20. **Bentrup, F. W.,** Vergleichende Untersuchungen zur Polaritätsinduktion durch das Licht an der *Equisetum*-spore und der *Fucus*zygote, *Planta,* 59, 472, 1963.
21. **Mosebach, G.,** Über den Einfluss des Lichtes auf die Polarisierung des befruchteten Eies von *Cystoseira barbata, Ber. Dtsch. Bot. Ges.,* 56, 210, 1938.
22. **Weber, W.,** Morphogenetische und keimungsphysiologische Untersuchungen an einigen Meeresalgen unter besonderer Berücksichtigung der Polarität, *Bot. Mar.,* 12, 135, 1969.
23. **Sand, O.,** On orientation of rhizoid outgrowth of *Ulva mutabilis* by applied electric fields, *Exp. Cell Res.,* 76, 444, 1973.
24. **Nultsch, W. and Pfau, J.,** Occurrence and biological role of light-induced chromatophore displacements in seaweeds, *Mar. Biol.,* 51, 77, 1979.
25. **Zurzycki, J.,** Blue light-induced intracellular movements, in *The Blue Light Syndrome,* Senger, H., Ed., Springer-Verlag, Berlin, 1980, 50.
26. **Nultsch, W., Pfau, J., and Ruffer, U.,** Do correlations exist between chromatophore arrangement and photosynthetic activity in seaweeds?, *Mar. Biol.,* 62, 111, 1981.
27. **Britz, S. J. and Briggs, W. R.,** Circadian rhythms of chloroplast orientation and photosynthetic capacity in *Ulva, Plant Physiol.,* 58, 22, 1976.
28. **Mizukami, M. and Wada, S.,** Action spectrum for light-induced chloroplast accumulation in a marine coenocytic green alga, *Bryopsis plumosa, Plant Cell Physiol.,* 22, 1245, 1981.
29. **Paques, M., Sironval, C., and Bonotto, S.,** On chloroplast movement in the stalk of *Acetabularia mediterranea,* in *Developmental Biology of Acetabularia,* Bonotto, S., Kefeli, V., and Puiseux-Dao, S., Eds., Elsevier/North Holland Biomedical Press, Amsterdam, 1979, 155.

30. **Paques, M. and Brouers, M.,** Chloroplast phototaxis in *Acetabularia mediterranea, Protoplasma,* 105, 360, 1981.

31. **Clauss, H.,** Effect of red and blue light on morphogenesis and metabolism of *Acetabularia mediterranea,* in *Biology of Acetabularia,* Brachet, J. and Bonotto, S., Eds., Academic Press, London, 1970, 177.

32. **Müller, S. and Clauss, H.,** Aspects of photomorphogenesis in the brown alga *Dictyota dichotoma, Z. Pflanzenphysiol.,* 78, 461, 1976.

33. **Velde, H. H. van der, Guiking, P., and Wulp, D. van der,** Glucose-6-phosphate dehydrogenase and 6-phosphogluconate dehydrogenase in *Acrochaetium daviesii* cultured under red, white and blue light, *Z. Pflanzenphysiol.,* 76, 95, 1975.

34. **Ruyters, G., Hirosawa, T., and Miyachi, S.,** Blue light effects on carbon metabolism in *Dunaliella,* in *Blue Light Effects in Biological Systems,* Senger, H., Ed., Springer-Verlag, Berlin, 1984, 317.

35. **Schmid, R.,** Blue light effects on morphogenesis and metabolism in *Acetabularia,* in *Blue Light Effects in Biological Systems,* Senger, H., Ed., Springer-Verlag, Berlin, 1984, 419.

36. **Kowallik, W. and Schürmann, R.,** Chlorophyll a/chlorophyll b ratios of *Chlorella vulgaris* in blue or red light, in *Blue Light Effects in Biological Systems,* Senger, H., Ed., Springer-Verlag, Berlin, 1984, 352.

37. **Humbeck, K. and Senger, H.,** The blue light factor in sun and shade plant adaptation, in *Blue Light Effects in Biological Systems,* Senger, H., Ed., Springer-Verlag, Berlin, 1984, 344.

38. **Lichtenthaler, H. K., Buschmann, C., and Rahmsdorf, U.,** The importance of blue light for the development of sun-type chloroplasts, in *The Blue Light Syndrome,* Senger, H., Ed., Springer-Verlag, Berlin, 1980, 485.

39. **Jeffrey, S. W.,** Responses of unicellular marine plants to natural blue-green light environments, in *Blue Light Effects in Biological Systems,* Senger, H., Ed., Springer-Verlag, Berlin, 1984, 497.

40. **Kamiya, A. and Miyachi, S.,** Blue-green and green light adaptations on photosynthetic activity in some algae collected from subsurface chlorophyll layer in the western Pacific Ocean, in *Blue Light Effects in Biological Systems,* Senger, H., Ed., Springer-Verlag, Berlin, 1984, 517.

41. **Faust, M. A., Sager, J. C., and Meeson, B. W.,** Response of *Prorocentrum mariae-lebouriae* (Dinophyceae) to light of different spectral qualities and irradiances: growth and pigmentation, *J. Phycol.,* 18, 349, 1982.

42. **Beer, S. and Levy, I.,** Effects of photon fluence rate and light spectrum composition on growth, photosynthesis and pigment relations in *Gracilaria* sp., *J. Phycol.,* 19, 516, 1983.

43. **Vesk, M. and Jeffrey, S. W.,** Effects of blue-green light on photosynthetic pigments and chloroplast structure in unicellular marine algae from six classes, *J. Phycol.,* 13, 280, 1977.

44. **Bogorad, L.,** Phycobiliproteins and complementary chromatic adaptation, *Annu. Rev. Plant Physiol.,* 26, 369, 1975.

45. **Marsac, N. T. de,** Occurrence and nature of chromatic adaptation in cyanobacteria, *J. Bacteriol.,* 130, 82, 1977.

46. **Ramus, J.,** A physiological test of the theory of complementary chromatic adaptation. II. Brown, green and red seaweeds, *J. Phycol.,* 19, 173, 1983.

47. **Dring, M. J. and Lüning, K.,** Induction of two-dimensional growth and hair formation by blue light in the brown alga *Scytosiphon lomentaria, Z. Pflanzenphysiol.,* 75, 107, 1975.

48. **Oltmanns, F.,** Über die Kultur- und Lebensbedingungen der Meeresalgen, *Jahrb. Wiss. Bot.,* 23, 349, 1892.

49. **Zumft, W. G., Castillo, F., and Hartmann, K. M.,** Flavin-mediated photoreduction of nitrate by nitrate reductase of higher plants and microorganisms, in *The Blue Light Syndrome,* Senger, H., Ed., Springer-Verlag, Berlin, 1980, 422.

50. **Terborgh, J.,** Effects of red and blue light on the growth and morphogenesis of *Acetabularia crenulata, Nature (London),* 207, 1360, 1965.

51. **Clauss, H.,** Beeinflussung der Morphogenese, Substanzproduktion und Proteinzunahme von *Acetabularia mediterranea* durch sichtbare Strahlung, *Protoplasma,* 65, 49, 1968.

52. **Lüning, K. and Dring, M. J.,** Reproduction, growth and photosynthesis of gametophytes of *Laminaria saccharina* grown in blue and red light, *Mar. Biol.,* 29, 195, 1975.

53. **Lüning, K.,** Critical levels of light and temperature regulating the gametogenesis of three *Laminaria* spp. (Phaeophyceae), *J. Phycol.,* 16, 1, 1980.

54. **Lüning, K. and Neushul, M.,** Light and temperature demands for growth and reproduction of laminarian gametophytes in Southern and Central California, *Mar. Biol.,* 45, 297, 1978.

55. **Lüning, K.,** Egg release in gametophytes of *Laminaria saccharina:* induction by darkness and inhibition by blue light and U.V., *Br. Phycol. J.,* 16, 379, 1981.

56. **Kumke, J.,** Beitrage zur Periodizität der Oogon-Entleerung bei *Dictyota dichotoma* (Phaeophyta), *Z. Pflanzenphysiol.,* 70, 191, 1973.

57. **Veilhaben, V.,** Zur Deutung des semilunaren Fortpflanzungszyklus von *Dictyota dichotoma, Z. Bot.,* 51, 156, 1963.

58. **Vince-Prue, D.,** *Photoperiodism in Plants,* McGraw-Hill, London, 1975.

59. **Dring, M. J.,** Phytochrome in red alga, *Porphyra tenera, Nature (London),* 215, 1411, 1967.
60. **Rentschler,H-G.,** Photoperiodische Induktion der Monosporenbildung bei *Porphyra tenera* Kjellm. (Rhodophyta-Bangiophyceae), *Planta,* 76, 65, 1967.
61. **Dring, M. J. and Lüning, K.,** A photoperiodic effect mediated by blue light in the brown alga *Scytosiphon lomentaria, Planta,* 125, 25, 1975.
62. **Hoopen, ten A.,** Regulatie van de vorming van tetrasporangia in het roodwier *Acrosymphyton purpuriferum* (J. Ag.) Sjöst, *Vakbl. Biol.,* 61, 364, 1981.
63. **Hoopen, ten A.,** Effects of Daylength and Irradiance on the Formation of Reproductive Organs in Two Algae: *Acrosymphyton purpuriferum* (J. Ag.) Sjöst. (Rhodophyceae) and *Sphacelaria rigidula* Kutz. (Phaeophyceae), Ph.D. thesis, University of Groningen, Groningen, Netherlands, 1983.
64. **Dring, M. J. and West, J. A.,** Photoperiodic control of tetrasporangium formation in the red alga *Rhodochorton purpureum, Planta,* 159, 143, 1983.
65. **Terry, L. A. and Moss, B. L.,** The effect of photoperiod on receptacle initiation in *Ascophyllum nodosum* (L.) Le Jol., *Br. Phycol. J.,* 15, 291, 1980.
66. **Senger, H.,** Cryptochrome, some terminological thoughts, in *Blue Light Effects in Biological Systems,* Senger, H., Ed., Springer-Verlag, Berlin, 1984, 72.
67. **Shropshire, W.,** Stimulus perception, in *Encyclopedia of Plant Physiology,* New Series, Vol. 7, *Physiology of Movements,* Haupt, W. and Feinleib, M. E., Ed., Springer-Verlag, Berlin, 1979, 10.
68. **Lüning, K.,** Tauchuntersuchungen zur Vertikalverteilung der sublitoralen Helgoländer Algenvegetation, *Helgol. Wiss. Meeresunters.,* 21, 271, 1970.
69. **Maggs, C. A.,** personal communication, 1985.
70. **Smith, H. and Morgan, D. C.,** The function of phytochrome in nature, in *Encyclopedia of Plant Physiology,* New Series, Vol. 16B, *Photomorphogenesis,* Shropshire, W. and Mohr, H., Eds., Springer-Verlag, Berlin, 1983, 491.
71. **Dring, M. J.,** Light quality and the photomorphogenesis of algae in marine environments, in *4th European Marine Biology Symposium,* Crisp, D. J., Ed., Cambridge University Press, Cambridge, 1971, 375.

Chapter 10

SUN AND SHADE EFFECTS OF BLUE LIGHT ON PLANTS

Horst Senger

TABLE OF CONTENTS

I.   Introduction 142

II.   Adaptation to Light Intensity 142
     A.   Higher Plants 142
     B.   Green Algae 142

III.   Adaptation to Light Quality 143
     A.   Higher Plants 143
     B.   Green Algae 143
     C.   Biosynthesis of Chlorophyll b and Light Harvesting
          Complex 145

IV.   Photoreceptors and Kinetics for Adaptation to Light Quality 147

V.   Summary 147

References 147

I. INTRODUCTION

Although all plant life depends on light, plants have only limited possibilities to choose the optimal light intensity and quality. They have to adapt their photosensitive systems, in particular the photosynthetic appartus, to optimize its reaction under the given, sometimes unfavorable, conditions.

Such adaptation can be a long evolutionary process which results in genetically fixed sun and shade plants with few possibilities for modification, or plants can have the flexibility to adapt to intensity and quality of light during their development.

The adaptation of the photosynthetic apparatus to different intensities of white light was reported by Willstätter and Stoll in 1913.[1] They detected a higher ratio of chlorophyl (Chl) a/b in sun leaves as compared to shade leaves from several plants. In past years these adaptation phenomena have been extensively studied for higher plants and to some extent for green algae. (Other groups of algae are considered in the preceding chapter.) Early attention has been paid to the fact that different light qualities cause effects similar to high and low light intensities. In 1936 Seybold and Egle[2] reported on the comparison of Chl a/b ratios of plants grown in "blue shade" and "green shade" with sun plants; but it has only been in recent years that the problem of adaptation to various light qualities has been pursued again. Studies on the influence of various wavelengths of light on the adaptation of the photosynthetic apparatus are more complicated.

So far little attention has been paid to questions about the nature of the photoreceptors for adaptation of the photosynthetic apparatus and how fast plants can adapt to a specific light condition.

II. ADAPTATION TO LIGHT INTENSITY

A. Higher Plants

Adaptation of higher plants to high and low light intensities has been widely studied and the results have been reviewed extensively.[3-5] Additionally, a detailed study reports on the light-intensity preference of photosynthetic aquatic organisms.[15] Resulting from adaptation to high and low intensities are the so-called sun and shade plants. Most plants are adaptable to either condition, like *Atriplex patula*[6] or *Sinapis alba*.[7] Other plants from extreme biotopes are genetically fixed as either sun plants, like *Tedestromia oblongifolia*,[8] or shade plants, like *Alocasia macrorhiza*.[8]

Characteristically, sunplants have a higher photosynthetic capacity than shade plants and reach compensation of photosynthesis at higher intensities and have higher respiration rates. For better utilization of light at low intensities, shade plants have more chlorophyll, thylakoids are orgainzed in grana, the chlorophyll a/b ratio is lower, and the ratio of chlorophyll to cytochrome f is higher than in sun plants. No significant differences have been detected in the quantum yield of oxygen evolution and the ratio of total chlorohyll and reaction centers. Changes in pigment protein complexes of the thylakoid membrane under varying light intensities have been recently reported.[9]

B. Green Algae

In a few green algae the adaptation of the photosynthetic apparatus to high and low intensities has been studied.[10-14,40] The results are very similar to those obtained for higher plants. Although no grana are formed, under low light intensities thylakoids demonstrate considerable stacking.

III. ADAPTATION TO LIGHT QUALITY

A. Higher Plants

Recent investigations on adaptation of the photosynthetic apparatus to red or blue light suffer from two exerimental difficulties: large fields of monochromatic blue light are not available, and the intensity factor is not considered. Light described as "blue light" is of wide spectral range, and some emission in the red can hardly be excluded. In various studies the intensities applied sometimes differ by as much as a factor of 10. Quite frequently (see Table 1) light intensities of wide spectral ranges or even of white light have been expressed as quantum flux in mol quanta $\cdot$ m^{-2} $\cdot$ sec^{-1}. These values can only be approximations since "mol quanta" are defined only for single wavelengths.

Nevertheless, comparing the scanty data available from the literature (Table 1) there are many discrepancies but some common features seem to emerge. Varying from experiment to experiment the amount of chlorophyll formed can be either higher in blue or in red light-grown plants. This might derive from the fact that the curves for chlorophyll formation during growth under different light qualities are almost identical in the beginning,[22,24] and that light intensity curves cross each other at low light intensities.[25] The pronounced difference in the Chl a/b ratio characterizing sun and shade plants is usually higher in the blue than in the red (Table 1). On the other hand, the ratio P 700/Chl appears to be higher in red light and there are more grana formed.

Regarding the light-intensity curves of photosynthesis in barley leaves, plants grown under blue light exhibit the pattern of sun plants, and those grown under red light show the pattern of shade plants.[25]

Summarizing, higher plants grown in red light resemble shade plants and those grown in blue light resemble sun plants. Nevertheless, one has to keep in mind that different intensities have not been considered for the different spectral regions. The developmental stage of the plants might add an additional factor which influences the adaptation of the photosynthetic apparatus.

B. Green Algae

Specific investigations on the adaptation of the photosynthetic apparatus of green algae are reported for *Chlorella*[26-28] and *Scenedesmus*,[29,30] with additional singular findings on *Ankistrodesmus*,[31] *Dunaliella*,[32] and *Acetabularia*.[33]

When cultures of green algae grown under blue light are compared to those under red light, a significant stimulation of mass production[28,31,33,34] (Figure 1) and total chlorophyll formation[26,27,30,32] (Table 2) are found. For the deviation from these general findings in mass production[26] and chlorophyll formation,[28] an explanation is currently not available. The ratio of Chl a/b was found to be lower in blue light cells in two cases[27,30] but higher in another study.[28] Reports on photosynthetic capacity of cells grown under blue or red light are controversial. In several publications photosynthetic capacity is reported to be higher in cells grown under blue light than in those grown under red light,[27,28,31,33] whereas the light-intensity curve of photosynthesis as based on an equal amount of chlorophyll demonstrates a controversial result (Figure 2). Similar data could be deduced from fluorescence measurements with *Chlorella*.[27] Cells grown under blue light show higher fluorescence than cells grown under red light when computed on an equal chlorophyll base; accordingly, photosynthesis of blue light-grown cells should be lower than of red light-grown cells.

Of the structural changes, Jeffrey[35] writes: "The major effects of low-intensity blue-green light on unicellular marine algae are increased pigment synthesis, no significant changes in pigment ratios... and more thylakoids per chloroplast...." In opposition to this statement, in *Chlorella* more stacked thylakoids were formed under red light than under blue light.[28]

Summarizing the adaptational phenomena of green algae to blue light, there appears to

Table 1

COMPARISON OF VARIOUS PARAMETERS OF HIGHER PLANTS GROWN UNDER RED (R) OR BLUE (B) LIGHT

Organism	Light Quality R	Light Quality B	Total Chl (R B)	Chl a/b (R B)	Grana (R B)	PS Units PSI (R B)	PS Units PSII (R B)	PS Units P-700/Chl (R B)	Ref.
Asplenium australasium	TL 40 W/15; 50 µE m⁻² sec⁻¹	TL 40 W/18; 50 µE m⁻² sec⁻¹	∨	∧	=	∧	∨	∨	16
Pisum sativum	665 nm max; 15 µE m⁻² sec⁻¹	435 nm max; 6 µE m⁻² sec⁻¹		∧		∧	≥	∨	17
Atriplex triangularis	TL 40 W/15; 5O µE m⁻² sec⁻¹	TL 40 W/18; 50 µE m⁻² sec⁻¹	∨	∧	∧		∧	∨	18
Sinapis alba	TL 40 W/15; 33 µE m⁻² sec⁻¹	TL 40 W/18; 33 µE m⁻² sec⁻¹	∨	δ	δ				19
Hordeum vulgare	660 nm max; 5.5 µE m⁻² sec⁻¹	450 nm max; 5.5 µE m⁻² sec⁻¹	∨	∨	∧				20
Raphanus sativus	660 nm max; 5.5 µE m⁻² sec⁻²	450 nm max; 5.5 µE m⁻² sec⁻¹	∨	≋	∧				20
Hordeum vulgare	660 nm max; 5.5 µE m⁻² sec⁻¹	450 nm max; 5.5 µE m⁻² sec⁻¹	∨	∨	∧				21
	660 nm max; 5.5 µE m⁻² sec⁻¹	450 nm max; 5.5 µE m⁻² sec⁻¹	∧	∨	∧				22
Phaseolus vulgaris	TL 40 W/15; 20.5 µE m⁻² sec⁻¹	450 nm max; 12.5 µE m⁻² sec⁻¹	∧	∨	∨				23
	TL 40 W/15; 20.5 µE m⁻² sec⁻¹	450 nm max; 12.5 µE m⁻² sec⁻¹	∧	∨	∨				24
Hordeum vulgare	660 nm max; 0.17 µE m⁻² sec⁻¹	480 nm max; 0.17 µE m⁻² sec⁻¹	∨			∨	∨		25

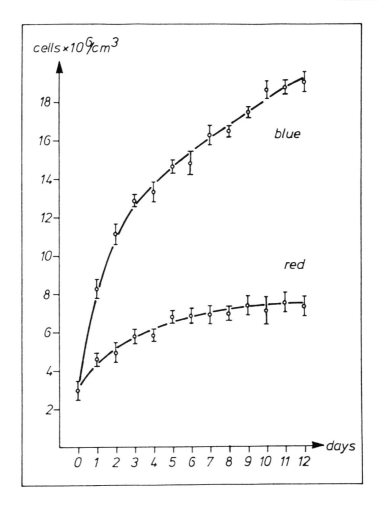

FIGURE 1.   Growth curves of cells of *Scenedesmus obliquus* grown under blue light (424 nm, 3,35 W · m^{-2}) and under red light (λ > 620 nm, 2.15 W · m^{-2}).

be a remarkable response of growth, pigment content, Chl a/b ratio, photosynthetic capacity, and chloroplast structure. Some results do not fit the general pattern, probably due to impurities of monochromatic light, different intensities, and neglect of developmental stages. At the current stage, it is not possible to match blue- or red-light-grown green algae unequivocally with shade and sun plants, but most results indicate that for green algae, blue light mimics shade conditions.

C. Biosynthesis of Chlorophyll b and Light-Harvesting Complex

In *Chlorella*[27] and *Scenedesmus*[30] the low ratio of Chl a/b already indicates a preferential formation of chlorophyll b under blue light. Chlorophyll b is only found in light-harvesting pigment complexes (LHC). A synchronous synthesis of LHC with chlorophyll b under blue light was in *Scenedesmus*.[30] This was supported by similar findings with cell suspensions of tobacco tissue.[37]

The formation of chlorophyll b and LHC is the response to shade conditions, evoked in higher plants by red light and in green algae by blue light. Again, tissue cultures of higher plants resemble the green algae more than the plants they are derived from.

Detailed studies of the blue light-dependent gene expression for the LHC proteins have been conducted with cell suspensions of tobacco.[36,37] The activity of mRNA corresponding

Table 2
CHLOROPHYLL CONTENT AND CHLOROPHYLL a/b RATIOS OF GREEN ALGAE GROWN UNDER BLUE (B) OR RED (R) LIGHT

Organism	Total Chl		Chl a/b		
	R	B	R	B	Ref.
Chlorella vulgaris	5.73	7.77[a]	5	3	27
Chlorella vulgaris	3.8	5.9[b]	—	—	26
Scenedesmus obliquus	9.3	13.8[c]	5.6	3.1	30
Chlorella fusca	15.2	9.7[d]	3.9	4.0	28
Dunaliella tertiolecta	0.94[e]	1.31[f]	—	—	32

[a]   Micrograms of chlorophyll per milligram of dry weight.
[b]   Relative units of chlorophyll per dry mass.
[c]   Migrograms of chlorophyll per microliter of PVC.
[d]   Milligrams of chlorophyll per gram of dry weight.
[e]   White light.
[f]   Micrograms of chlorophyll per 10^6 cells.

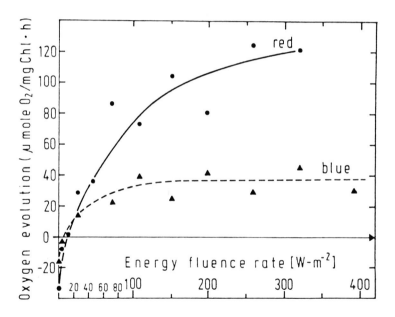

FIGURE 2.   Photosynthetic oxygen evolution of cells of *Scenedesmus obliquus* grown under blue and red light as indicated in Figure 1. Oxygen evolution was measured under red light ($\lambda > 620$ nm of various intensities) with an oxygen electrode.

to the synthesis of LHC proteins was measured using cloned chloroplast genes from *Sinapis* and cDNA constructed from (poly A$^+$-RNA from tobacco as probes. The data show that blue light-induced changes in mRNA activity correspond to changes in the level of the mRNA sequence. Blue light seems to induce the activity of mRNA in the nucleus and chloroplast, i.e., a transient increase in the rate of transcription. Experiments with an active transcription complex isolated from greening tobacco cells supported this notion.

IV. PHOTORECEPTORS AND KINETICS FOR ADAPTATION TO LIGHT QUALITY

Plants capable of adapting to light quantities and qualities will adjust their photosynthetic apparatus to "shade conditions" when transferred to red light (higher plants), or to blue light (green algae, tissue cultures). These adaptations are regulated by the light qualities. The acting photoreceptors are most probably not photosynthetic pigments. Although detailed investigations on this topic are still pending, it is generally assumed that phytochrome regulates this adaptation of the photosynthetic apparatus in higher plants.[4] The fact that green algae show many responses to blue light which are not inducible in red light makes it most probable that a blue light photoreceptor regulates the adaptation of the photosynthetic apparatus to light qualities.

The difficulty in studying action spectra for the transformation of sun into shade plants or vice versa stems from the rather long adaptation times. For the transfer of *Sinapis* from shade- to sun-type chloroplasts, 5 days were necessary.[38] For the transformation of red light-adapted chloroplasts to blue light-adapted chloroplasts, *Acetabularia* needed 72 hr.[39] Recent studies conducted in our laboratory indicated that the transformation of strong- and weak-light chloroplasts into weak- and strong-light chloroplasts, respectively, took only 6 hr in homocontinuous cultures of *Scenedesmus*. Therefore, unicellular green algae seem to be a good tool for determining action spectra for adaptation in the future.

V. SUMMARY

Adaptations of the photosynthetic apparatus to high and low intensities of white light take place in a similar fashion in higher plants and in green algae. Application of blue and red light points in the following direction: for higher plants, red light simulates shade conditions and for green algae and tissue cultures of higher plants blue light induces adaptation phenomena like shade conditions. Modifications of these general statements might be necessary when intensities of the different light qualities and developmental stages are considered.

The response of higher plants and green algae to red and blue light, respectively, seems to be in accordance with ecological situations. Under the canopy of plants, the blue and the red part of the spectrum are filtered away, whereas green and far-red light penetrate.[41] In clear waters the long-wavelength part of the spectrum disappears whereas blue-green light can penetrate to a depth of 200 m.[42] It is quite logical that higher plants and algae growing in their specific habitat are using photoreceptors for adaptation sensitive to the specific light qualities available.

REFERENCES

1. **Willstätter, R. and Stoll, A.,** *Untersuchungen über Chlorophyll,* Springer-Verlag, Berlin, 1913.
2. **Seybold, A. and Egle, K.,** Lichtfeld und Blattfarbstoffe, *Planta,* 26, 491, 1937.
3. **Boardman, N. K.,** Comparative photosynthesis of sun and shade plants, *Annu. Rev. Plant Physiol.,* 29, 355, 1977.
4. **Wild, A.,** Physiologie der Photosynthese Höherer Pflanzen. Die Anpassung an die Lichtbedingungen, *Ber. Dtsch. Bot. Ges.,* 92, 341, 1979.
5. **Björkman, O.,** Responses to different quantum flux densities, in *Encyclopedia of Plant Physiology,* New Series, Vol. 12A, *Responses to Physical Environment,* Lange, O. L., Nobel, P. S., Osmond, C. B., and Ziegler, H., Eds., Springer-Verlag, Berlin, 1981, 57.

6. **Boradman, N. K., Björkmann, O., Anderson, J. M., Goodchild, D. J., and Thorne, S. W.,** Photosynthetic adaptation of higher plants to light intensity: relationship between chloroplast structure, composition of the photosystems and photosynthetic rates, in *Proc. 3rd Int. Congr. Photosynthesis Res.,* Avron, M., Ed., Elsevier, Biomedical Press, Amsterdam, 1974, 1809.

7. **Grahl, H. and Wild, A.,** Studies on the content of P-700 and cytochromes in *Sinapis alba* during growth under two different light intensities, in *Environmental and Biological Control of Photosynthesis,* Marcelle, R., Ed., Dr. W. Junk Publishers, The Hague, 1975, 107.

8. **Berry, J. A.,** Adaptation of photosynthetic processes to stress, *Science,* 188, 644, 1975.

9. **Leong, T.-Y. and Andersen, J. M.,** Adaptation of the thylakoid membranes of pea chloroplasts to light intensities. I. Studies on the distribution of chlorophyll-protein complexes, in *Photosynthesis Research,* Vol. V, Sybesma, C., Ed., Martinus Nijhoff/Dr. W. Junk Publishers, The Hague, 1984, 105.

10. **Noddack, W. and Eichhoff, H. I.,** Untersuchungen über die Assimilation der Kohlensäure durch die grünen Pflanzen. II. Assimilation und Lichtintensität, *Z. Phys. Chem.,* 185, 222, 1939.

11. **Steeman Nielsen, E.,** Inactivation of the photochemical mechanism in photosynthesis as a means to protect the cells against too high light intensities, *Physiol. Plant.,* 15, 161, 1962.

12. **Reger, J. B. and Krauss, R. W.,** The photosynthetic response to a shift in the chlorophyll a to chlorophyll b ratio in *Chlorella, Plant Physiol.,* 46, 568, 1970.

13. **Senger, H. and Fleischhacker, Ph.,** Adaptation of the photosynthetic apparatus of *Scenedesmus obliquus* to strong and weak light conditions. I. Differences in pigments, photosynthetic capacity, quantum yield and dark reactions, *Physiol. Plant.,* 43, 35, 1978.

14. **Fleischhacker, Ph. and Senger, H.,** Adaptation of the photosynthetic apparatus of *Scenedesmus obliquus* to strong and weak light conditions. II. Differences in photochemical reactions, the photosynthetic electron transport and photosynthetic units, *Physiol. Plant.,* 43, 43, 1978.

15. **Phlips, E. J. and Mitsui, A.,** Light intensity preference and tolerance of aquatic photosynthetic microorganisms, in *Handbook of Biosolar Resources,* Vol. I, Mitsui, A. and Black, C. C., Eds., CRC Press, Boca Raton, Fla., 1982, 257.

16. **Leong, T.-Y., Goodchild, D. J., and Anderson, J. M.,** Effect of light quality on the composition, function, and structure of photosynthetic thylakoid membranes of *Asplenium australasicum* (Sm.) Hook, *Plant Physiol.,* 78, 561, 1985.

17. **Leong, T.-Y. and Anderson, J. M.,** Changes in composition and function of thylakoid membranes as a result of photosynthetic adaptation of chloroplasts from pea plants grown under different light conditions, *Biochim. Biophys. Acta,* 723, 391, 1983.

18. **Leong, T.-Y. and Anderson, J. M.,** Effect of light quality on the composition and function of thylakoid membranes in *Atriplex triangularis, Biochim. Biophys. Acta,* 766, 533, 1984.

19. **Wild, A. and Holzapfel, A.,** The effect of blue and red light on the content of chlorophyll, cytochrome f, soluble reducing sugars, soluble proteins and the nitrate reductase activity during growth of the primary leaves of *Sinapis alba,* in *The Blue Light Syndrome,* Senger, H., Ed., Springer-Verlag, Berlin, 1980, 444.

20. **Lichtenthaler, H. and Buschmann, C.,** Control of chloroplast development by red light, blue light and phytohormones, in *Developments in Plant Biology,* Vol. 2, *Chloroplast Development,* Akoyunoglou, G. and Argyroudi-Akoyunoglou, J. H., Elsevier/North-Holland, Biomedical Press, Amsterdam, 1978, 801.

21. **Buschmann, C., Meier, D., Kleudgen, H. K., and Lichtenthaler, H. K.,** Regulation of chloroplast development by red and blue light, *Photochem. Photobiol.,* 27, 195, 1978.

22. **Lichtenthaler, H. K., Buschmann, C., and Rahmsdorf, U.,** The importance of blue light for the development of sun-type chloroplasts, in *The Blue Light Syndrome,* Senger, H., Ed., Springer-Verlag, Berlin, 1980, 485.

23. **Anni, H. and Akoyunoglou, G.,** The effect of blue and red light on the development of the photosynthetic units during greening of etiolated bean leaves, in *Photosynthesis,* Vol. V, *Chloroplast Development,* Akoyunoglou, G., Ed., Balaban International Science Services, Philadelphia, 1981, 885.

24. **Akoyunoglou, G. and Anni, H.,** Blue light effect on chloroplast development in higher plants, in *Blue Light Effects in Biological Systems,* Senger, H., Ed., Springer-Verlag, Berlin, 1984, 397.

25. **Voskresenskaya, N. P.,** Control of the activity of photosynthetic apparatus in higher plants, in *Blue Light Effects in Biological Systems,* Senger, H., Ed., Springer-Verlag, Berlin, 1984, 407.

26. **Kubin, S., Borns, E., Doucha, J., and Seiss, U.,** Light absorption and production rate of *Chlorella vulgaris,* in light of different spectral composition, *Biochem. Physiol. Pflanz.,* 178, 193, 1983.

27. **Kowallik, W. and Schürmann, R.,** Chlorophyll a/Chlorophyll b ratios of *Chlorella vulgaris* in blue or red light, in *Blue Light Effects in Biological Systems,* Senger, H., Ed., Springer-Verlag, Berlin, 1984, 352.

28. **Wilhelm, Ch., Krämer, P., and Wild, A.,** Effect of different light qualities on the ultrastructure, thylakoid membrane composition and assimilation metabolism of *Chlorella fusca, Physiol. Plant.,* 64, 359, 1985.

29. **Humbeck, K. and Senger, H.,** The blue light factor in sun and shade plant adaptation, in *Blue Light Effects in Biological Systems,* Senger, H., Ed., Springer-Verlag, Berlin, 1984, 344.

30. **Humbeck, K., Schumann, R., and Senger, H.,** The influence of blue light on the formation of chlorophyll-protein complexes in *Scenedesmus,* in *Blue Light Effects in Biological Systems,* Senger, H., Ed., Springer-Verlag, Berlin, 1984, 359.

31. **Adler, K.,** Spezifische Rolle der Carotinoidabsorption bei der photosynthetischen Sauerstoffentwicklung, *Planta,* 75, 220, 1967.

32. **Vesk, M. and Jeffrey, S. W.,** Effect of blue-green light on photosynthetic pigments and chloroplast structure in unicellular marine algae from six classes, *J. Phycol.,* 13, 280, 1977.

33. **Clauss, H. and Schael, U.,** Die Wirkung von Rot- und Blaulicht auf die Photosynthese von *Acetabularia mediterranea, Planta,* 78, 98, 1968.

34. **Humphrey, G. F.,** The effect of spectral composition of light on the growth, pigments and photosynthetic rate of unicellular marine algae, *J. Exp. Mar. Biol. Ecol.,* 66, 49, 1983.

35. **Jeffrey, S. W.,** Responses of unicellular marine plants to natural blue-green light environments, in *Blue Light Effects in Biological Systems,* Senger, H., Ed., Springer-Verlag, Berlin, 1984, 497.

36. **Richter, G.,** Blue light control of the level of two plastid mRNAs in cultured plant cells, *Plant Mol. Biol.,* 3, 271, 1984.

37. **Richter, G. and Wessel, K.,** Red light inhibits blue light-induced chloroplast development in cultured plant cells at the mRNA level, *Plant Mol. Biol.,* 5, 175, 1985.

38. **Grahl, H. and Wild, A.,** Lichtinduzierte Veränderungen im Photosynthese-Apparat von *Sinapis alba, Ber. Dtsch. Bot. Ges.,* 86, 341, 1973.

39. **Schmid, R.,** Blue light effects on morphogenesis and metabolism in *Acetabularia,* in *Blue Light Effects in Biological Systems,* Senger, H., Ed., Springer-Verlag, Berlin, 1984, 319.

40. **Reger, J. and Krauss, R. W.,** The photosynthetic response to a shift in the chlorophyll a to chlorophyll b ratio of *Chlorella, Plant Physiol.,* 46, 568, 1970.

41. **Holmes, M. G. and Smith, H.,** The function of phytochrome in the natural environment. II. The influence of vegetation canopies on the spectral energy distribution of natural daylight, *Photochem. Photobiol.,* 25, 539, 1977.

42. **Jerlov, N. G.,** *Marine Optics,* Elsevier Oceanographic Series, Volume 14, Elsevier, Amsterdam, 1976.

Index

INDEX

A

Absorption spectrum, see also Action spectroscopy;
    Spectrophotometry; Spectroscopy
  flavins, 4—6
  membrane fractions, 94
Accessory pigments
  action spectra of, 46—47
  marine plants, 128
Acetabularia
  action spectra, 133, 134
  chromatophore displacement, 126
  enzymes controlled by blue light, 74
  growth in red light, 128, 147
  growth regulation, 136, 137
  hair formation, 130
  light quality adaptations, 128, 143, 147
  morphogenesis, 58, 62, 63
  protein and carbohydrate metabolism, 127
  reproductive development, 130
Acetabularia mediterranea, 74
Acid-base properties, flavins, 20
Acifluorfen, 94
Acrochaetium, 127
Acrochaetium daviesii, 73, 74, 127
Acrosymphyton, 132—134
Actin-myosin cytoskeletal elements, 55, 60, 66
Actinomycin D, 56
Action spectroscopy
  action spectra, 40—45
  classical, 38
  complex photoreceptors, 46—49
  experimental approaches used with, 49—51
  physical principles, 38—40
  reciprocity, 43—44
  single photoreceptor systems, kinetic models for,
    45—46
Action spectrum, see also Action spectroscopy;
    Spectroscopy
  carotenogenesis in *Neurospora crassa*, 14
  cauliflower LIAC, 92
  chloroplast formation, 59
  marine plants, 133—134
  spectrophotometry, 30—31
  sun vs. shade effects, adaptation kinetics, 147
Adaptation kinetics, 147
Agmenellum quadruplicatum, 73
ALA dehydratase, 73
Alaria, 125
ALA synthase, 73
Alcohol dehydrogenase, 73
Aldolase, 73, 81
Algae, see also Marine plants; genera by name
  sun and shade effects, 142, 143, 145
Allomyces, 64
Alocasia macrorhiza, 142
Alternaria tomato, 49
Amino acid metabolism, 80

Amino acid oxidase, 78
AMP, 82
Amylase, 81
Anabaena flos-aquae, 74
Analogs, photoreceptor, 51
Ankistodesmus, 143
Antheridium formation, 130
Anthocyanins, 59, 96
Apoproteins, 13
Arabidopsis thaliana, 51
Ascophyllum, 126
Asplenium australasium, 144
ATP, 79
ATPases, 105—106, 118
Atriplex patula, 142
Atriplex triangularis, 144
Auxins, 7, 63—64
Avena, 63—66
Azide, 49, 94

B

Bacteriophages, 45
Bacteriorhodopsin, 11, 20, 21
Bean, 64, 113, 144
Behavioral mutants, 50—51
Bilirubin, 11
Biochemical development, 54—56
Bioluminescence reactions, 12
Blue-green algae, phototaxis, 125
Blue-green light, 123, 124
  effects on marine algae, 143
  phytoplankton responses, 137
Blue shade, 142
Broad bean, 113
Bryopsis, 126, 133
Bunsen-Roscoe law of reciprocity, 46

C

Calvin cycle, 74
cAMP phosphodiesterase, 73
Canopy, 147
Carbohydrate metabolism, see also Enzymes, con-
    trol mechanisms, 79—82, 127
Carbon dioxide, 117
Carbon monoxide difference spectra, 92
β-Carotene, 50—51
Carotenogenesis, 8
Carotenoids, 9—11
  analogs, 51
  exogenous, carotenoidless mutant effects, 12
  in *Neurospora*, 96
  role of, 12—14
  spectroscopy, 9—10
Catalase, 73

Catalysis, photochrome, 46—48
Cauliflower membrane purification, 92—96, 100—106
CCCP, 117
Cell cultures
   gene expression
      light-harvesting complex, 145—146
      morphogenesis, 56, 58
   light quality adaptations, 143
   morphogenesis, 54, 58, 59
   mRNA induction, 56
Cell membranes
   purification procedure, 100—106
   receptors, see Receptors, membrane-bound
Chitin synthase, 73
Chlamydomonas
   enzymes controlled by blue light, 73, 74
   mutants, 12, 50—51
   phototaxis, retinal analogs and, 51
Chlamydomonas reinhardii, 73, 74
Chlorella
   enzyme regulation in, 72—74, 77—82
   LIACs, 24
   morphogenesis, chloroplast formation, 59
   pigment content, 127
   sun/shade effects, 143, 145, 146
Chlorella ellipisoidea, 73
Chlorella pyrenoidosa, 73, 74, 146
Chlorogonium elongatum, 73, 74
Chlorophylls
   guard cell, 112
   marine plants, 127—128, 137
   sun/shade effects, light quality adaptation, 143, 145—146
Chloroplasts
   formation of, 57—59
   light quality effects, 142, 143
   sun/shade effects, light intensity adaptation, 142
   tobacco cell culture, 54
Chromatic transients, 47
Chromatophore displacement, marine plants, see also Photoreceptors, 126
Chrysazine, 12
Cinnamic acid-4-hydroxylase, 96
Circadian rhythms, marine plants, 126
Cis-trans isomerization, 11, 20—22
Citrate synthase, 81
Coccolithophorids, 125
Codium fragile, 126
Cofactors, and enzyme control, 72, 74, 78
Coherent-anti-Stokes Raman scattering, 22
Coleone A, 12
Commelina communis, 112—119
Competitive inhibition, 74
Complex photoreceptors, 46—49
Conductance, 112, 113, 116
Conformation, 21—23
Conidiation, 49, 55
Continuous light
   Acetabularia morphogenesis, 63
   in higher plants, 59

for potentiation, 54
Coomassie blue, 60
Corn, 73, 74
   flavin-cytochrome b electron transfer reaction, 8
   LIACs, 31
   methylene blue reaction, 90
   quenchers of excited states, 49
Cosmarium, 81
Criterion response, 41
Cryptochrome, see also Receptors, 31, 32, 59
Cryptomonads, 128
Cryptomonas, 125—126, 128
Cryptophyta, 125
Crysophyta, 125
Cucumber, 56, 73
Cucumis sativus, 56, 73
Cyanide-insensitive respiration, 64—65
Cyanidium caldarium, 74
Cyanobacteria, 63, 125
Cyanophyta, 125
Cycloheximide, 56
Cystathione, 8
Cystoseira, 126
Cytochalasin B, 66
Cytochrome, b-type, 77
   in *Dictyostelium,* 95
   electron transfer reaction, 8
   as LIAC, 30
   nitrate reductase, 23
   quantum yield, 90
Cytochrome b-557, 77
Cytochrome c, membrane markers, 105
Cytochrome c oxidase, 93
Cytochrome f, 142
Cytochrome oxidase, 46, 93
Cytochrome P-450, 92—93
Cytoplasm, changes in organization of, 65—66
Cytoskeleton
   electric fields and, 63
   morphogenesis, 55

D

Dark-grown organisms
   Chlorella mutants, enzyme regulation in, 79
   chloroplast formation, 59
   Euglena, chloroplast formation, 57—58
DCMU, 117
Deazaflavin, 6, 14
Decarboxylation, oxidative, 7
Deexcitation processes, 39—40
Degradation, flavins, 6
DES, and stomatal response to blue light, 117
Developmental responses, marine plants, 128—132
Dextran, 91, 100—105
Dichroism, 5, 49
Dichromatic irradiation, 48, 49
Dictyostelium discoideum, 54, 56, 90, 95
Dictyota, 127, 130—134
   growth in red light, 128, 129

growth regulation, 136, 137
    reproductive development, 130
Differential absorbance spectra, 94
Dihydroflavin, 32
Dihydroflavylium salts, 12
7,8-Dimethyl-9-(formylmethyl)isoalloxazine, 6
Dinoflagellates, 125
Dipole moments, 21, 22
Dipole transition moment, 49—50
Dissociation reactions, 20—22
Dithionite, 92, 94
Diuron, 59
DNA, see also Transcription/translation
    Neurospora transformation, 66
DNA photolyase, 13
DNA repair, 12
Dolichos lablab, 73, 74
DOVA dehydrogenase, 73
DOVA transaminase, 73
Dual-wavelength spectrophotometer, 26, 27, 30—
    31, 90
Dunaliella, 125, 133
    light quality adaptations, 143
    protein and carbohydrate metabolism, 127
Duration of light exposure, see Pulse duration
Dyes
    photoresponses with, 7, 12
    reduction of, by hyphal tips, 64

E

EDTA, 7, 23
Electrical changes, 55, 62—65
Electronic state assignments, carotenoids, 9, 10
Electron transfer reactions
    flavin-mediated, 8, 9
    cytochrome P-450, 93
Electrophoresis, 58, 60, 61, 92
Elodea, 65
Endogenous rhythms, 132—133
Energy level diagram, flavins, 5
Energy transfer reactions, 21
Enolase, 81
Enzymes
    control mechanisms, 72, 74—76
        coarse control, 76—77
        direct action of blue light, 77—78
        fine control, 77—83
        indirect action of blue light, 78—83
    flavin, blue light effects, 23—24
    marine plants, 127
    in morphogenesis, 54, 59
    photoinactivation of, 45
    photoreactivating, in bacteria, 14
    range affected by blue light, 72—74
Eosin, 7
Erythrosin, 7
Escherichia coli, 14
Ethionine, 8
Ethylene, 8

Etiolation, rapid effects in, 56—57
Euglena, 8, 57—59
Euglena gracilis, 6, 73
Excited states
    quenchers of, 49
    photon absorption and, 38—39
    relaxation of photoreceptor pigments, 20—22
Eximers, 21, 22
Exiplexes, 21, 22

F

FAD, 23, 24, 77, 93
Far-red light
    canopy effects, 147
    marine plants, 137
    photosynthetic effectiveness of, 46—47
    phytochrome action spectra, 47
FbP, 82, 83
Feedback inhibition, 74
Fern, 59—60, 65
Flavins, 4—8
    acid-base properties, 20
    analogs of, 51
    cytochrome complexes, 92—96
    dipole transition moments, 49, 50
    fluorescence and phosphorescence emissions, 6
    inhibitors of, 27
    photoreactivities of, 12
    quenchers, 49
    redox state, 22—23
    role of, 14
    spectroscopy, 4—6
Flavonoid photoprotectants, UV-B induction of, 54
Flavoproteins, and respiratory pathways, 64—65
Flavosemiquinone, 32
Fluence rate, conversion to photon fluence rate, 38
Fluence-response curves, measurement of action
        spectra without, 44—45
Fluorescence, 6, 20, 39
Fluorescence polarization, 26
FMN, 5, 23, 24, 77, 93
Forbidden transitions, 27, 39
Fucoxanthin, 11, 128
Fucus, 126, 133, 134
Fumarase, 73, 81
Fusarium aqueductuum, 90

G

Gas exchange, 112—119
Gel electrophoresis, 58, 60, 61, 92
Gel phase, lipids, 31—32
Genetics, 51
Gene transcription, see also Transcription/translation
    induction of, 54
    light-harvesting complex, 145—146
Germ tube development, 126
Gilvin, 122

Glucan synthase II, 106
β-D-Glucose oxidase, 23
Glucose oxidase, 78, 91
Glucose-6-phosphate, 82, 83
Glucose-6-phosphate dehydrogenase, 73, 81, 127
Glutamate dehydrogenase, 81
Glutamate oxaloacetate dehydrogenase, 81
Glutamate pyruvate transaminase, 81
Glutamate synthase, 72, 73, 81
Glyceraldehyde-3-phosphate dehydrogenase, 73, 74,
        78, 81
Glycine max, 113
Glycine oxidase, 23, 77, 78
Glycolate oxidase, 73, 78
Glyoxylate oxidase, 73
GOGAT, 72, 73
Gonyaulax, 125
Gracilaria, 128
Green algae, 126
Greening systems, 57—59, 66
Green light, 136—137, 147
Green shade, 142
Griffithsia, 125
Grotthus-Draper law, 20, 38
Ground state, 38—40
Growth regulation, marine plants, 135—137
Guard cells, proton extrusion from protoplasts, 117,
        118
Gymnodinium, 125, 133

H

Hair formation, marine algae, 130, 135—136
Halobacterium, 12—14, 51
Halobacterium halobium, 4
Hartmann model, 47—48
HeLa cells, 31, 90
Hexokinase, 78, 81, 83
Homocystine, 8
Hordeum, 74
Hordeum vulgare, 74, 144
Hydrogen peroxide, 24, 32
Hydroxypyruvate reductase, 73
Hyphae
    growth in electric fields, 63
    reduction of indicator dyes, 64

I

IAA, 7
Inhibitors
    of chloroplast formation, 59
    flavin, 27
    membrane studies, 94, 95
    protein synthesis, 77
    transcription/translation, 56
In vitro studies
    cell culture, see Cell culture
    enzyme effects of blue light, 77

Iodine, 11, 14
Ionic currents, 55
Irradiance response curves, 59
Isoalloxazine nucleus, 6
Isocitratase, 73
Isocitrate dehydrogenase, 81
Isocoleone A, 12
Isoenzymes, *Chlorella* mutants, 79
Isomerization, 14, 21

J

Jablonski diagram, 38, 39

K

α-Ketoglutarate dehydrogenase, 73, 81
Kinetic photochrome model, 47—48
Konig-Kramers transformation, 5—6

L

Lactate dehydrogenase, 78, 81
Laminaria, 130—131, 133—137
Laminaria saccharina, 129, 130
Laminariales, 123
LIACs, see Light-induced absorbance changes
Light exposure, duration of, see Pulse duration
Light-harvesting complex, 145—146
Light-induced absorbance changes (LIACs), 27, 28,
        30
    diversity of, 95—96
    electron transfer reaction, 8
    flavins in, 14
    as marker, 105
    membrane purification, 92—94
    number of, 91—94
    relevance of, 31—32, 90—91
Light intensity, see also Photon fluence
    adaptation to, 142
    marine plant responses, 134—135
    sun vs. shade effects, 147
Light-intensity gradients, 13
Light quality, see also Wavelength
    marine plant responses, 134—135
    sun and shade effects, 142—148
    of underwater light, 122—124
Lipids, phase transitions, 32
Lithium dodecylsulfate-PAGE gels, 92
Lumichrome, 6
Luminescence reactions, 10, 12

M

Macrocystis, 134
Macrocystis pyrifera, 130
Malate dehydrogenase, 78, 81
Malonate, 74
Marine plants
    developmental responses, 128—132
    environment of, 122—124
    growth regulation, 135—137
    light quality effects, 143
    metabolic responses, 127—128
    photoorientation responses, 124—126
    photoperiodic responses, 132—133
    physiological aspects of responses, 133—135
Markers, plasma membrane, 105—106
Mass action ratio, 76, 79
Membrane potential, changes in, 64
Membranes
    purification procedure, 100—106
    receptors, see Receptors, membrane-bound
Messenger RNA, see Transcription/translation
Metabolic responses, marine plants, 127—128
Methionine, 8
Methylene blue, 7, 64, 90
Microcystis aeruginosa, 74
Microtubules, 55, 60, 66
Molecular orbital calculations, flavins, 5
Molybdenum cofactor, 23, 77
Morphogenesis, 54—66
    biochemical development, 54—56
    early effects, 54
    rapid effects, 56—57, 62—66
    transcription/translation, 54, 56, 57—62
    transduction chain, 54
Mosses, 126
mRNA, see RNA, messenger; Transcription/
        translation
Munoz-Butler system, see LIACs
Mycochrome, 49

N

NADH, 7
NAD kinase, 73
NAD(P), 82
NADP-cytochrome c reductase, 93
NAD(P)H, 23
NADP protochlorophyllide oxidoreductase, 73
Naphthazarins, 12
Naphthyl phthalamic acid, 91
Near UV, 64
Neurospora
    carotenoid biosynthesis, 96
    LIACs, 24, 31—32
    morphogenesis, 54, 58, 64, 66
    nitrate reductase, 95
Neurospora crassa, 90
    carotenoidless mutant, 12

enzymes controlled by blue light, 73
    LIACs in, 27—31
    methylene blue reaction, 91
Nicotiana tabacum, see also Tobacco cell culture,
        73, 74
Night breaks, 132
Nitrate reductase, 23, 31, 73, 78, 95
    control mechanisms, 77
    photocatalysis by, 46
Nitrite reductase, 73
NMR, 66, 79—83
Nucleic acids, see Transcription/translation

O

Oat, 63—66
Ochromonas, 127
Onoclea, 65
Orchid, 113
Oxidation
    of auxins, 7
    carotenoids, 14
    flavin, 6, 7
Oxidation-reduction reactions
    enzyme control via, 74, 75
    relaxation of excited pigment, 21
Oxidation state, flavin, 22—23
Oxidative decarboxylation, 7
2'-Oxoflavin, 6
2-Oxoglutarate dehydrogenase, 72
Oxygen
    carotenoid isomerization and, 11
    carotenoid photooxidation and, 14
    evolution of red vs. blue light effects, 146
    molecular, 7
Oxygen tension, and methylene blue reaction, 91

P

Paphiopedilum harrisianum, 113
Paramecium, 7, 12
Pariser-Parr-Popple, 6
Parsley, cell culture, 54, 56, 58
Pelvetia, 126
Percoll gradient, 93
Peridinin, 13, 128
Peridinium, 125
Peroxides, 24, 32
Petalonia, 129
pH, and stomatal response to blue light, 117
Phages, 45
Phalloidin, 66
Pharbitis nil, 73
Phaseolus, 73, 74
Phaseolus vulgaris, 64, 113, 144
Phenylacetic acid, 94
Phenylalanine ammonia lyase, 73
Phormidium, 63, 125
Phosphoenolpyruvate (PEP), 79, 82

Phosphoenolpyruvate (PEP) carboxylase, 73, 81, 79
Phosphofructokinase, 74, 77, 78, 81, 83
6-Phosphogluconate dehydrogenase, 74, 81, 127
Phosphoglycerate kinase, 74
Phosphorescence, 20, 39
Phosphorylase, 81
Photoacoustic spectroscopy, 25
Photobilirubin, 11
Photocatalysis, 46
Photochromes, 47—48
Photoinactivation, 45
Photon absorption, 38—39
Photon fluence, 38, 119
    action spectra measurement without fluence re-
        sponse curves, 44—45
    marine plant responses, 134—135
    photochrome catalysis, 48
    pulse duration and, 43
    response as function of, 40—43
    sun and shade effects, 142—148
Photoorientation responses, marine plants, 125—126
Photoperiod, 132—134
Photophobic response, 7
Photoreceptors
    guard cell, 112
    primary, 4—14
        carotenoids, 9—11
        flavins, 4—8
        other chromophores, 11—12
        speculative remarks, 12—14
    primary reactions, 20—24
    spectrophotometry, 24—32
Photorespiration, see also Enzymes, control mecha-
    nisms; Respiration
Photoreversibility, of photoresponse, 48—49
Photosensitizing dyes, see Dyes
Photosynthetic effectiveness, of far-red light, 46—
    47
Phototaxis, 8
    marine plants, 125—126, 133—134
    retinal analogs and, 51
Phototropism
    electrical changes in, 63—64
    electric fields and, 63—64
    marine plants, 125
    Phycomyces blakesleeanus, multiple receptors, 11
    quenchers and, 49
Phycobilins, 128
Phycocyanin, 128
Phycoerythrin, 125, 128
Phycomyces, 24
    action spectra in, 44
    electric fields and, 64
    flavin singlet-triplet transition, 27
    morphogenesis, 58, 62
    mutants, behavioral, 50
    near-red action peak in, 12
    rapid effects in, 56, 57
    transition dipole moments in, 49
Phycomyces blakesleeanus, 11, 73, 90, 96
Physarum morphogenesis

time course of, 55
    spherulation and sporulation, 60, 61
    transcriptional/translational events, 58
Physiology, marine plants, 133—135
Phytochrome
    action spectra of, 47
    guard cell, 112
    marine plants, 137
Phytoplankton, see also Marine plants, 123, 137
Pigment, marine plants, 127—128
Pisum, 73, 74
Pisum arvense, 73
Pisum sativum, 74, 144
Plasma membrane, see Membranes; Receptors,
    membrane-bound
Plastids, 57—59
Platymonas, 125, 133
Polarity, induction of, 126
Polarizability, 5, 21, 22, 49—50
Polarization, fluorescence, 26
Polarized single-crystal absorption, 5
trans-Polyenes, 10
Polyethylene glycol, 91, 100—105
Polysiphonia, 125
Porphyra tenera, 132
Porphyria, 125
Porphyridium cruentum, 125
Potassium iodide, 94
Prasinophyta, 125
Primary photoreceptors, see Photoreceptors, primary
Primary reactions, see also Photoreceptors
    morphogenetic
        cytoplasm, changes in organization of, 65—66
        electrical changes, 62—65
        transduction chain, 54
    rates of, in action spectroscopy, 40
Product formation, 45, 72, 74, 75
Prorocentrum, 125, 128
Protein metabolism, see also Enzymes, control
        mechanisms; Transcription/translation
    enzyme control mechanism, 77
    marine plants, 127
Protochlorophyllide, 57, 58
Proton exchange, 21
    extrusion from guard cell protoplast, 117, 118
    flavin-mediated, 8
Pteridiophytes, 126
Pulse duration, 43
    morphogenesis and, 54
    stomatal response to blue light, 114—116
Puromycin, 56
Pyridine-binding spectra, 92
Pyruvate dehydrogenase, 81
Pyruvate kinase, 72, 74, 77—79, 81, 82, 127

Q

Quantum efficiency/yield, 38, 40
    cytochrome b reduction, 90
    of phototransformation, 42—43

Quantum flux, sun and shade effects, 142—148
Quantum mechanically-forbidden transitions, 27, 39, 49
Quarter wave stack, 13
Quasicrystalline layer, carotenoids as, 13
Quenching, 49

R

Radiative transfer, 39—40
Raphanus sativus, 144
Rate constants, phototransformation, 38, 40
Reaction rates, in action spectroscopy, 40
Receptors, membrane-bound
  inhibitor studies, 94, 95
  LIACs
    diversity of, 95—96
    number of, 91—94
    physiological relevance of, 90—91
    purification procedures, 91
Reciprocity, 43—44
Red algae, 125, 128, 134
Red drop, 46
Red light
  higher plants grown in, 143, 144
  marine plant responses, 128—129, 137
  phytochrome action spectra, 47
  sun vs. shade effects, 143—147
Redox reactions
  enzyme control via, 74, 75
  relaxation of excited pigment, 21
  in sensory transduction chain, 27
Redox-state, flavin, 22—23
Reducing agents, hyphal tips as, 64
Reduction, of membrane fractions, 93
Reference wavelength, 42
Renografin centrifugation, 91
Reproductive development, marine plants, 130—132
Resonance Raman spectroscopy, 10, 22
Respiration, see also Enzymes, control mechanisms
  in *Chlorella,* 24
  cyanide-insensitive, and morphogenesis, 64—65
Retinal, 12, 22, 51
Reversibility, of photoresponse, 48—49
Rhizoid development, 126
Rhodochorton, 134
Rhodophyta, 125, 126
Rhodopsin, 11
  in *Chlamydomonas,* 12, 51
  relaxation of excited pigment, 20, 21
  retinal and, 51
Rhodopsin-like receptors, 4, 12, 14, 51
Riboflavin, 24, 49—51
Ribulose-1,5-bisphosphate carboxylase, 59, 74, 77, 78
RNA, messenger, see also Transcription/translation
  light-harvesting complex, 145—146
  in morphogenesis, 54
  in parsley cell cultures, 56
  photoregulation of, 77

RNA polymerase, 74
Roseoflavin, 51
Rubisco, 59, 74, 77, 78

S

Saccharomyces cerevisiae, 14
Saccharum, 113
Salicylhydroxamic acid (SHAM), 64—65
Scenedesmus, 73, 74
  enzyme regulation in, 79, 81—82
  morphogenesis, 59, 66
  pigment content, 127
  sun vs. shade effects, 143, 145—147
Scenedesmus obliquus, 73, 74, 146
Schiff's base, 11, 12, 21, 22
Schizophyllum, 63
Scytosiphon, 125, 130, 133—135
  action spectra, 133
  growth regulation, 136, 137
  hair formation, 130
  phototropism, 125
Scytosiphon lomentaria, 129, 132, 136
Sensory transduction, see Transduction
Sieve effect, 25
Silicoflagellates, 125
Silicotungstic acid, 91
Sinapis, 147
Sinapis alba, 73, 142, 147
Single-beam spectrophotometer, 26, 27, 29—30
Single-crystal absorption, FMN, 5
Single-photoreceptor systems, 45—46
Singlet state
  flavins, 6—8, 27, 32
  oxygen, 7, 24, 32
  return to ground states, 39—40
Singlet-triplet transition, 27, 39, 49
Slime mold, see specific genera; specific species
Soybean, 113
Spectral distribution, see Wavelength
Spectrophotometry, 24—32
Spectroscopy
  action, see Action spectroscopy
  carotenoids, 9—10
  flavins, 4—6
  low-temperature, 93
  photoacoustic, 25
Spirodela, 58, 59
Sporangiophores, initiation of, 57
Sporulation
  Physarium, 60
  time course, 55
  UV-B control of, 54
Stark-Einstein law of equivalence, 38
Starvation, and blue light effects, 24
Stephanoptera, 125
Stomatal repsonse, 111—119
  kinetic model, 114—117
  kinetic properties, 112—114

mechanistic and functional implications, 117—
119
proton extrusion from guard cell protoplast, 117,
118
Streptomyces griseus, 14
Stress, and induction, 57
Substrate, enzyme regulation, 72, 74, 75, 78
Succinate dehydrogenase, 74, 81
Sugarcane, 113
Sun and shade effects
light intensity, adaptation to, 142
light quality, adaptation to, 143—146
algae, 143, 145
chlorophyll b biosynthesis, 145—146
higher plants, 143, 144
photoreceptors for and kinetics of, 147—148
Sunflower, 56
Superoxide, 24, 32

T

Tedestromia oblongifolia, 142
Tobacco, 73, 74
Tobacco cell culture
chloroplast formation, 54
enzymes controlled by blue light, 73, 74
gene expression, light-harvesting complex, 145—
146
morphogenesis, 58, 59
Transcription, 8
Transcription/translation, see also Enzymes, control
mechanisms
chloroplast formation, 54—57
control mechanisms, 62, 63
light harvesting complexes, red vs. blue light effects, 145—146
without morphogenesis, 54, 56
morphogenesis, external structures, 59—62
morphogenesis, internal, 54—57
posttranslational events, enzyme control via, 74
Transduction
general scheme for, 5
morphogenetic processes, 54
redox reaction in, 27
relaxation of excited photoreceptor pigments,
20—22
Transition dipole moments, 5—6, 49
Transketolase, 78
Trans-polyenes, 10
Trichoderma, 14
conidiation, 60—62
duration of light exposure for, 54

electric fields in, 64
time course of development, 55
transcriptional/translational events, 58
Triplet state, flavin, 6, 7, 32
acid-base properties, 20
forbidden transitions, 27, 49
intersystem crossing to, 39
Triticum aestivum, 73
Tubulins, 60
Turbidity, and light absorption, 24—26

U

UDPG, 82
UDPG pyrophosphorylase, 62, 63, 74, 127
Ulva, 125, 133
Ulva latuca, 126
Ulva mutabilis, 126
Uricase, 78
UV-light, 54, 64

V

Vaucheria morphogenesis, 55, 64, 66
Vegetative development, marine plants, 129—130
Velocity, maximal, 76, 82
Vicia faba, 73, 74, 113, 118
Viruses, photoinactivation of, 45
Volvocales, 125

W

Wavelength
chloroplast formation and, 59
photon-influence rate response curves, 40—42
photoreversibility of response, 48—49
quantum efficiency and, 38, 40
underwater light, 122—124, 135
Wolffia arrhiza, 73

X

Xanthine oxidase, 78

Z

Zea mays, see also Corn, 73, 74